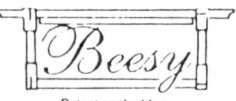

THE
BEEKEEPERS ANNUAL
1996

The Beekeepers' Annual is published by:
Northern Bee Books
Scout Bottom Farm
Mytholmroyd
Hebden Bridge
West Yorks.   HX7 5JS
Tel: (01422) 882751   Fax: (01422) 886157

and edited by
Ron Coles
Manor Farm
Upton, Nr. Newark
Notts. NG23 5ST
Tel: (01636) 812289

Designed by
K. Sutcliffe

Cover
*Clarice Cliff Beehive Honeypot - **Liberty***

2

CONTENTS

FOREWORD	Ken Ibbotson	4
EDITORIAL	Ron Coles	5
CALENDAR AND DIARY		7
HONEYPOTS, MONEYPOTS	Leonard Griffin	61
O, TO BE IN ENGLAND	George Davis	63
BEES CAUGHT IN THE NET	John Richards	65
NOT A LOT OF PEOPLE KNOW THAT	Brian Dennis	68
CARRYING ON THE WORK	Ron Coles	73
CROSSWORD	Byron	75
BILL GETS ELECTRIFIED	Bill Clark	80
HAVING A FIELD DAY	Paul Smith	86
JUMP ON YOUR CAMEL	John Kinross	88
C.B. DENNIS	David Little	91
CROSSWORD SOLUTION		92
DIRECTORY OF BEEKEEPING ORGANISATIONS AND STATISTICS		95

FOREWORD

by Ken Ibbotson

THE BEEKEEPERS' ANNUAL goes from strength to strength. It is not only a mine of current information in each of its publication years, but through its diary section it allows each beekeeper to compile a valuable personal record within a contemporary context. It is likely that some collections of BKA will become valuable historical documents; if only we had had such a publication during the days of the big beemasters of the past!

On looking through past copies one is struck by the range of topics covered. There is something of value for the small honey producer, including the true amateur and/or hobby beekeeper, through to the largest of commercial enterprises. As beekeeping is affected more and more by uncertain shifting sands as the international politicians and the Commissioners in Brussels increase some regulations on the one hand, and 'deregulate' on the other, it is vital to have a reliable current reference source to which one can turn.

Although European and international negotiations move at a snail's pace, decisions do emerge. The GATT agreement of last year included a reduction on the tariff on honey into the European Union over the next six years. This will increase imports from countries with lesser overheads at a time when costs at home continue to rise and new regulations add to those costs. Many beekeepers are putting pressure on MPs, MEPs, MAFF and the Commission to provide a 'level playing field', for home honey producers and to recognise the value of beekeeping to the ecological and environmental structure of Europe.

The Beekeepers' Annual is a valuable publication to illustrate the strength of the beekeeping industry in the British Isles. Long may it continue and we are indebted to the editor and his team who compile such a worthwhile contribution to the beekeeping scene.

Ken Ibbotson, Hon. Sec. CONBA

EDITORIAL

 S I PUT the finishing touches to the 1996 edition of the Beekeepers' Annual, the weather has turned cold, wet and windy. It seems hardly possible that only a few short weeks ago we were wondering if the hottest summer in 200 years would ever break. Actually this year's heat wave was down to me. Earlier this year we decided to splash out and have professionals in to landscape the garden, the full works, not just the layout but hundreds of container grown plants, designed to provide colour all the year round and forage for the bees. The last time we went into anything like this was in a previous house and the year was 1976—need I say more. Actually we fared rather better this year and most of the plants were nursed through the drought. Fortunately, the day after the hosepipe ban came into force the weather broke and the rains came down.

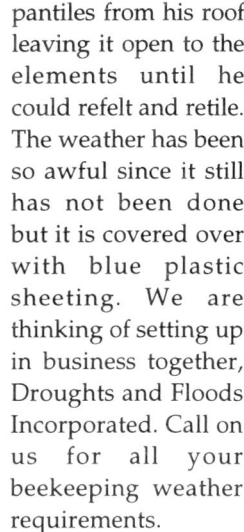

That was down to my next door neighbour. This was the day he chose to strip all the pantiles from his roof leaving it open to the elements until he could refelt and retile. The weather has been so awful since it still has not been done but it is covered over with blue plastic sheeting. We are thinking of setting up in business together, Droughts and Floods Incorporated. Call on us for all your beekeeping weather requirements.

I hope 1996 will provide the perfect climatic mix for bumper honey production and until the season starts I hope you will enjoy this edition of the Beekeepers' Annual.

Ron Coles

THE
BEEKEEPERS ANNUAL
1996

Calendar & Diary 1996

Illustrations from
Buzz a Buzz or The Bees
by W.C.C.
*taken from **Schnurrdiburr** by William Busch*

CALENDAR 1996

JANUARY
M	T	W	T	F	S	S
1	2	3	4	5	6	7
8	9	10	11	12	13	14
15	16	17	18	19	20	21
22	23	24	25	26	27	28
29	30	31				

FEBRUARY
M	T	W	T	F	S	S
			1	2	3	4
5	6	7	8	9	10	11
12	13	14	15	16	17	18
19	20	21	22	23	24	25
26	27	28	29			

MARCH
M	T	W	T	F	S	S
				1	2	3
4	5	6	7	8	9	10
11	12	13	14	15	16	17
18	19	20	21	22	23	24
25	26	27	28	29	30	31

APRIL
M	T	W	T	F	S	S
1	2	3	4	5	6	7
8	9	10	11	12	13	14
15	16	17	18	19	20	21
22	23	24	25	26	27	28
29	30					

MAY
M	T	W	T	F	S	S
		1	2	3	4	5
6	7	8	9	10	11	12
13	14	15	16	17	18	19
20	21	22	23	24	25	26
27	28	29	30	31		

JUNE
M	T	W	T	F	S	S
					1	2
3	4	5	6	7	8	9
10	11	12	13	14	15	16
17	18	19	20	21	22	23
24	25	26	27	28	29	30

JULY
M	T	W	T	F	S	S
1	2	3	4	5		
6	7	8	9	10	11	12
13	14	15	16	17	18	19
20	21	22	23	24	25	26
27	28	29	30	31		

AUGUST
M	T	W	T	F	S	S
			1	2	3	4
5	6	7	8	9	10	11
12	13	14	15	16	17	18
19	20	21	22	23	24	25
26	27	28	29	30	31	

SEPTEMBER
M	T	W	T	F	S	S
						1
2	3	4	5	6	7	8
9	10	11	12	13	14	15
16	17	18	19	20	21	22
23	24	25	26	27	28	29
30						

OCTOBER
M	T	W	T	F	S	S
	1	2	3	4	5	6
7	8	9	10	11	12	13
14	15	16	17	18	19	20
21	22	23	24	25	26	27
28	29	30	31			

NOVEMBER
M	T	W	T	F	S	S
				1	2	3
4	5	6	7	8	9	10
11	12	13	14	15	16	17
18	19	20	21	22	23	24
25	26	27	28	29	30	

DECEMBER
M	T	W	T	F	S	S
						1
2	3	4	5	6	7	8
9	10	11	12	13	14	15
16	17	18	19	20	21	22
23	24	25	26	27	28	29
30	31					

JANUARY

At early dawn 'tis quite a treat
To see them work, they are so neat;
Some clean their house with broom and mops,
And others empty out the slops.

January 1996

1 *Mon*	**9** *Tues*
2 *Tues*	**10** *Wed*
3 *Wed*	**11** *Thurs*
4 *Thurs*	**12** *Fri*
5 *Fri*	**13** *Sat*
6 *Sat*	**14** *Sun*
7 *Sun*	**15** *Mon*
8 *Mon*	**16** *Tues*

17 *Wed*	**25** *Thurs*
18 *Thurs*	**26** *Fri*
19 *Fri*	**27** *Sat*
20 *Sat*	**28** *Sun*
21 *Sun*	**29** *Mon*
22 *Mon*	**30** *Tues*
23 *Tues*	**31** *Wed*
24 *Wed*	

Day	Forage	Weather						Hive Weights		
		Temp.		Wind		Cloud Cover	Rain Fall	1	2	3
		Min	Max	DIR	B.S.					
1										
2										
3										
4										
5										
6										
7										
8										
9										
10										
11										
12										
13										
14										
15										
16										
17										
18										
19										
20										
21										
22										
23										
24										
25										
26										
27										
28										
29										
30										
31										

FEBRUARY

The architects, by rule and line,
Their future cells with skill refine;
The ever toiling workers these —
Meanwhile the Queen, she takes her ease;
Sole mother of the winged nation,
Her only work is propagation.

February 1996

1 *Thurs*	**9** *Fri*
2 *Fri*	**10** *Sat*
3 *Sat*	**11** *Sun*
4 *Sun*	**12** *Mon*
5 *Mon*	**13** *Tues*
6 *Tues*	**14** *Wed*
7 *Wed*	**15** *Thurs*
8 *Thurs*	**16** *Fri*

17 *Sat*	25 *Sun*
18 *Sun*	26 *Mon*
19 *Mon*	27 *Tues*
20 *Tues*	28 *Wed*
21 *Wed*	29 *Thurs*
22 *Thurs*	
23 *Fri*	
24 *Sat*	

Day	Forage	Weather						Hive Weights		
		Temp.		Wind		Cloud Cover	Rain Fall	1	2	3
		Min	Max	DIR	B.S.					
1										
2										
3										
4										
5										
6										
7										
8										
9										
10										
11										
12										
13										
14										
15										
16										
17										
18										
19										
20										
21										
22										
23										
24										
25										
26										
27										
28										
29										
30										
31										

MARCH

The egg she lays; the nurses hatch
That egg, and in the cradle watch.
The babe to swaddle, and prepare
The pap-boat, is their constant care.

March 1996

1 *Fri*	**9** *Sat*
2 *Sat*	**10** *Sun*
3 *Sun*	**11** *Mon*
4 *Mon*	**12** *Tues*
5 *Tues*	**13** *Wed*
6 *Wed*	**14** *Thurs*
7 *Thurs*	**15** *Fri*
8 *Fri*	**16** *Sat*

17 *Sun*	**25** *Mon*
18 *Mon*	**26** *Tues*
19 *Tues*	**27** *Wed*
20 *Wed*	**28** *Thurs*
21 *Thurs*	**29** *Fri*
22 *Fri*	**30** *Sat*
23 *Sat*	**31** *Sun*
24 *Sun*	

Day	Forage	Weather									Hive Weights		
		Temp.		Wind		Cloud Cover		Rain Fall			1	2	3
		Min	Max	DIR	B.S.								
1													
2													
3													
4													
5													
6													
7													
8													
9													
10													
11													
12													
13													
14													
15													
16													
17													
18													
19													
20													
21													
22													
23													
24													
25													
26													
27													
28													
29													
30													
31													

All day, in regal state, the Queen
Encircled by her court is seen;
Their backs they never rudely turn:
Good manners they by instinct learn.

April 1996

1 Mon	**9** Tues
2 Tues	**10** Wed
3 Wed	**11** Thurs
4 Thurs	**12** Fri
5 Fri	**13** Sat
6 Sat	**14** Sun
7 Sun	**15** Mon
8 Mon	**16** Tues

17 *Wed*	**25** *Thurs*
18 *Thurs*	**26** *Fri*
19 *Fri*	**27** *Sat*
20 *Sat*	**28** *Sun*
21 *Sun*	**29** *Mon*
22 *Mon*	**30** *Tues*
23 *Tues*	
24 *Wed*	

Day	Forage	Weather						Hive Weights		
		Temp.		Wind		Cloud Cover	Rain Fall	1	2	3
		Min	Max	DIR	B.S.					
1										
2										
3										
4										
5										
6										
7										
8										
9										
10										
11										
12										
13										
14										
15										
16										
17										
18										
19										
20										
21										
22										
23										
24										
25										
26										
27										
28										
29										
30										
31										

The air is clear the day is warm,
John Dull sits watching for a swarm.

May 1996

1 *Wed*	**9** *Thurs*
2 *Thurs*	**10** *Fri*
3 *Fri*	**11** *Sat*
4 *Sat*	**12** *Sun*
5 *Sun*	**13** *Mon*
6 *Mon*	**14** *Tues*
7 *Tues*	**15** *Wed*
8 *Wed*	**16** *Thurs*

17 *Fri*	**25** *Sat*
18 *Sat*	**26** *Sun*
19 *Sun*	**27** *Mon*
20 *Mon*	**28** *Tues*
21 *Tues*	**29** *Wed*
22 *Wed*	**30** *Thurs*
23 *Thurs*	**31** *Fri*
24 *Fri*	

Day	Forage	Weather								Hive Weights		
		Temp.		Wind		Cloud Cover		Rain Fall		1	2	3
		Min	Max	DIR	B.S.							
1												
2												
3												
4												
5												
6												
7												
8												
9												
10												
11												
12												
13												
14												
15												
16												
17												
18												
19												
20												
21												
22												
23												
24												
25												
26												
27												
28												
29												
30												
31												

John Dull, awakened from his slumber,
Observed his stock's diminished number;
His apple trees he searched and found
The swarm some ten feet from the ground

June 1996

1 *Sat*	**9** *Sun*
2 *Sun*	**10** *Mon*
3 *Mon*	**11** *Tues*
4 *Tues*	**12** *Wed*
5 *Wed*	**13** *Thurs*
6 *Thurs*	**14** *Fri*
7 *Fri*	**15** *Sat*
8 *Sat*	**16** *Sun*

17 *Mon*	25 *Tues*
18 *Tues*	26 *Wed*
19 *Wed*	27 *Thurs*
20 *Thurs*	28 *Fri*
21 *Fri*	29 *Sat*
22 *Sat*	30 *Sun*
23 *Sun*	
24 *Mon*	

Day	Forage	Weather							Hive Weights		
		Temp.		Wind		Cloud Cover		Rain Fall			
		Min	Max	DIR	B.S.				1	2	3
1											
2											
3											
4											
5											
6											
7											
8											
9											
10											
11											
12											
13											
14											
15											
16											
17											
18											
19											
20											
21											
22											
23											
24											
25											
26											
27											
28											
29											
30											
31											

Got his bee dress, his hive, and ladder;
No Bee master was ever gladder.

July 1996

1 Mon	**9** Tues
2 Tues	**10** Wed
3 Wed	**11** Thurs
4 Thurs	**12** Fri
5 Fri	**13** Sat
6 Sat	**14** Sun
7 Sun	**15** Mon
8 Mon	**16** Tues

17 *Wed*	**25** *Thurs*
18 *Thurs*	**26** *Fri*
19 *Fri*	**27** *Sat*
20 *Sat*	**28** *Sun*
21 *Sun*	**29** *Mon*
22 *Mon*	**30** *Tues*
23 *Tues*	**31** *Wed*
24 *Wed*	

Day	Forage	Weather								Hive Weights		
		Temp.		Wind		Cloud Cover		Rain Fall		1	2	3
		Min	Max	DIR	B.S.							
1												
2												
3												
4												
5												
6												
7												
8												
9												
10												
11												
12												
13												
14												
15												
16												
17												
18												
19												
20												
21												
22												
23												
24												
25												
26												
27												
28												
29												
30												
31												

Mounted, and without any trip
Got all the bees within the skip —

August 1996

1 *Thurs*	**9** *Fri*
2 *Fri*	**10** *Sat*
3 *Sat*	**11** *Sun*
4 *Sun*	**12** *Mon*
5 *Mon*	**13** *Tues*
6 *Tues*	**14** *Wed*
7 *Wed*	**15** *Thurs*
8 *Thurs*	**16** *Fri*

17 *Sat*	**25** *Sun*
18 *Sun*	**26** *Mon*
19 *Mon*	**27** *Tues*
20 *Tues*	**28** *Wed*
21 *Wes*	**29** *Thurs*
22 *Thurs*	**30** *Fri*
23 *Fri*	**31** *Sat*
24 *Sat*	

Day	Forage	Weather						Hive Weights		
		Temp.		Wind		Cloud Cover	Rain Fall	1	2	3
		Min	Max	DIR	B.S.					
1										
2										
3										
4										
5										
6										
7										
8										
9										
10										
11										
12										
13										
14										
15										
16										
17										
18										
19										
20										
21										
22										
23										
24										
25										
26										
27										
28										
29										
30										
31										

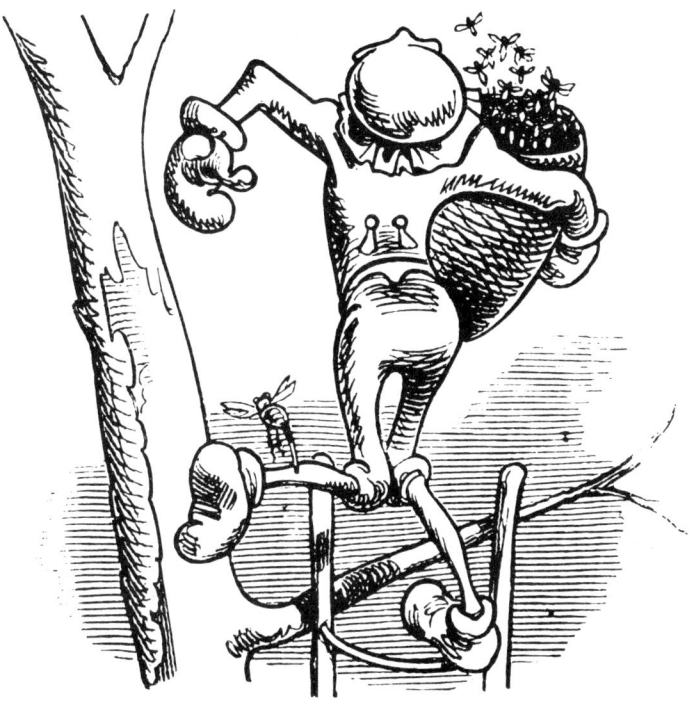

"Well done I have them;" as he spoke
The ladder's top-most rung it broke

September 1996

1 *Sun*	**9** *Mon*
2 *Mon*	**10** *Tues*
3 *Tues*	**11** *Wed*
4 *Wed*	**12** *Thurs*
5 *Thurs*	**13** *Fri*
6 *Fri*	**14** *Sat*
7 *Sat*	**15** *Sun*
8 *Sun*	**16** *Mon*

17 *Tues*	25 *Wed*
18 *Wed*	26 *Thurs*
19 *Thurs*	27 *Fri*
20 *Fri*	28 *Sat*
21 *Sat*	29 *Sun*
22 *Sun*	30 *Mon*
23 *Mon*	
24 *Tues*	

Day	Forage	Weather						Hive Weights		
		Temp.		Wind		Cloud Cover	Rain Fall	1	2	3
		Min	Max	DIR	B.S.					
1										
2										
3										
4										
5										
6										
7										
8										
9										
10										
11										
12										
13										
14										
15										
16										
17										
18										
19										
20										
21										
22										
23										
24										
25										
26										
27										
28										
29										
30										
31										

Crack! Crack! and, as I hope to thrive,
The same befel the other five

October 1996

1 *Tues*	**9** *Wed*
2 *Wed*	**10** *Thurs*
3 *Thurs*	**11** *Fri*
4 *Fri*	**12** *Sat*
5 *Sat*	**13** *Sun*
6 *Sun*	**14** *Mon*
7 *Mon*	**15** *Tues*
8 *Tues*	**16** *Wed*

17 *Thurs*	**25** *Fri*
18 *Fri*	**26** *Sat*
19 *Sat*	**27** *Sun*
20 *Sun*	**28** *Mon*
21 *Mon*	**29** *Tues*
22 *Tues*	**30** *Wed*
23 *Wed*	**31** *Thurs*
24 *Thurs*	

Day	Forage	Weather								Hive Weights		
		Temp.		Wind		Cloud Cover		Rain Fall		1	2	3
		Min	Max	DIR	B.S.							
1												
2												
3												
4												
5												
6												
7												
8												
9												
10												
11												
12												
13												
14												
15												
16												
17												
18												
19												
20												
21												
22												
23												
24												
25												
26												
27												
28												
29												
30												
31												

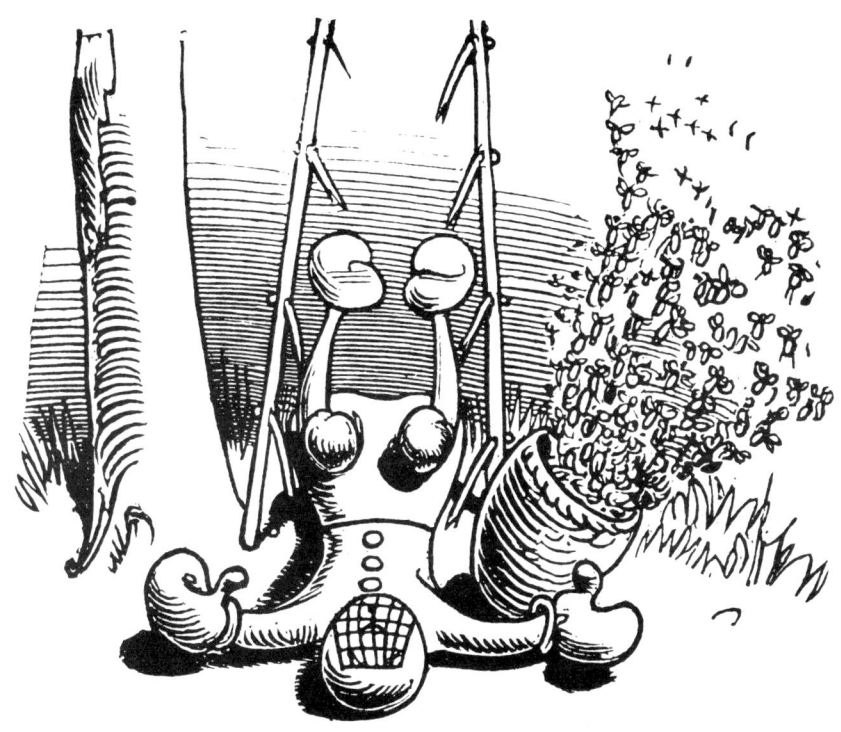

The bees rush forth and quit the hive!

November 1996

1 *Fri*	**9** *Sat*
2 *Sat*	**10** *Sun*
3 *Sun*	**11** *Mon*
4 *Mon*	**12** *Tues*
5 *Tues*	**13** *Wed*
6 *Wed*	**14** *Thurs*
7 *Thurs*	**15** *Fri*
8 *Fri*	**16** *Sat*

17 *Sun*	**25** *Mon*
18 *Mon*	**26** *Tues*
19 *Tues*	**27** *Wed*
20 *Wed*	**28** *Thurs*
21 *Thurs*	**29** *Fri*
22 *Fri*	**30** *Sat*
23 *Sat*	
24 *Sn*	

Day	Forage	Weather								Hive Weights		
		Temp.		Wind		Cloud Cover	Rain Fall			1	2	3
		Min	Max	DIR	B.S.							
1												
2												
3												
4												
5												
6												
7												
8												
9												
10												
11												
12												
13												
14												
15												
16												
17												
18												
19												
20												
21												
22												
23												
24												
25												
26												
27												
28												
29												
30												
31												

John on his knees, and free from harm
Marked well the dissappearing swarm.

December 1996

1 *Sun*	**9** *Mon*
2 *Mon*	**10** *Tues*
3 *Tues*	**11** *Wed*
4 *Wed*	**12** *Thurs*
5 *Thurs*	**13** *Fri*
6 *Fri*	**14** *Sat*
7 *Sat*	**15** *Sun*
8 *Sun*	**16** *Mon*

17 *Tues*	25 *Wed*
18 *Wed*	26 *Thurs*
19 *Thurs*	27 *Fri*
20 *Fri*	28 *Sat*
21 *Sat*	29 *Sun*
22 *Sun*	30 *Mon*
23 *Mon*	31 *Tues*
24 *Tues*	

Day	Forage	Weather						Hive Weights		
		Temp.		Wind		Cloud Cover	Rain Fall	1	2	3
		Min	Max	DIR	B.S.					
1										
2										
3										
4										
5										
6										
7										
8										
9										
10										
11										
12										
13										
14										
15										
16										
17										
18										
19										
20										
21										
22										
23										
24										
25										
26										
27										
28										
29										
30										
31										

HIVE RECORDS

Hive/ Q No.	Year Q Raised	Frames of Brood Autumn 1995	Combs Covered	Honey stored sugar fed Kg	Combs Covered Spring 1996	Frames of Brood Spring 1996	Spring feeding KG	Queens Reared	Nuclei
1									
2									
3									
4									
5									
6									
7									
8									
9									
10									
11									
12									
13									
14									
15									
16									
17									
18									
19									
20									
21									
22									
23									
24									

HONEYBEE COLONIES

1								
2								
3								
4								
5								
6								
7								
8								
9								
10								
11								
12								
13								
14								
15								
16								
17								
18								
19								
20								
21								
22								
23								
24								

BEEKEEPING RECORDS

Number	Items	Est. Value £	p
	Stocks of Bees		
	Empty Hives		
	Combs - Deep		
	Shallow		
	Frames		
	Foundation		
	Honey Extractor		
	Honey Tanks		
	Other Items		
	Honey Jars		
	Honey		
	Wax		
	Total		

HONEYPOTS, MONEYPOTS

by Leonard Griffin

HONEYPOTS IN the shape of beehives with a bee modelled as the handle are very traditional, but ironically the examples pictured are actually the work of Britain's leading *Art Deco* designer, Clarice Cliff. Born in Tunstall, Staffordshire in 1899 she was at first just an enameller and gilder for two adjoining factories, Wilkinson's and Newport Pottery. Until 1928 her ambition was to be a sculptress, and to this end she was the factory's modeller from the middle of the Twenties. The pieces she produced at this time included cartoonish duck figures marketed as bookends or an egg cup set, a *Viking Boat* flower-holder and her *Beehive Honeypot*. Her aim was to model shapes to amuse or entertain, as Britain was in the middle of the Depression and inventive ware was essential to capture sales.

Her role at the factories changed suddenly when she left behind the Victorian taste that still dominated the Potteries in the Twenties. In l928 she discovered Art Deco design and used this new style for pottery—cubist teapots with triangular handles, oblong plates and geometric vases. She became a huge success and by 1930 was made Art Director. It was unheard of for a woman to achieve this post in the Staffordshire "pot banks" at the time.

However, Clarice had a strong love of flowers and natural shapes so alongside

*Clarice Cliff **Beehive Honeypots**—left an example in **Sungold** from 1934, and right, one in a design called **Liberty** from 1930. Note the small "door" on the right hand honeypot!*

the Art Deco ware she continued to produce *fancies* such as the *Beehive Honeypot*. Clarice did not dream up the idea of the shape, we can trace this back as far as the middle of the Victorian era when examples were produced in silver by Elkington of Birmingham. One might be tempted to believe Clarice had first designed it, however, as between 1927 and 1939 she sold thousands of examples.

Her beehive was highly modelled including a small door and a big bumble bee, with its wings spread, sitting on the side of the hive. Over sixty years after the shape first appeared it is one of the most collectible of her *fancies* because as was the case with all her shapes it was available

From the original 1928 pattern books, Clarice's hand-written instructions to her paintresses as to how to paint the bee on a particular pot!

in any of several hundred hand painted designs! Although by 1930 many British factories were producing ware with printed decoration, Clarice Cliff cultivated a special decorating shop to hand paint her wares. She achieved this by taking young girls straight from school aged 14. They had nimble fingers capable of executing her designs onto the pottery at speed and were eager to work for the small but regular wages! As her pottery was sold under the banner of *Bizarre* ware, they inevitably became known as the "*Bizarre girls*". The sixty paintresses were soon extremely talented and learned how to fit Clarice's

designs onto tall candlesticks or low shapes such as the Beehive Honeypot. The legacy for collectors now is the chance to collect a shape they can find in many of these designs, and as it was made in two sizes a group of these can be very attractive.

Clarice Cliff's ware first became collectible during her lifetime and a group of enthusiasts held an exhibition at Brighton Museum shortly before she died in 1972. Since then many books and exhibitions have stimulated a mass of interest and raised the value of major pieces of her work. Initially collectors were able to buy the honeypots for as little as £10 to £20, but competition amongst collectors meant the more desirable or rare designs started to sell for between £150 and £300! Large vases and plaques cost anything from £500 to £7000, and Christie's of South Kensington now hold two **all** Clarice Cliff auctions annually! All this amazes the surviving Clarice Cliff paintresses who are now 77 to 85 years old, as when first sold the larger size Beehive Honeypot cost just seventeen shillings (eighty-five pence) for a dozen!

(Leonard Griffin is Chairman of the Clarice Cliff Collectors Club and author of "Clarice Cliff the Bizarre Affair" published by Thames & Hudson)

O, TO BE IN ENGLAND

by George Davis

O, TO BE IN England now that April's here" is an old saying. After more than sixty years of beekeeping, I am looking forward to seeing this time again.

My first hives were made when I was an apprentice, that is, when I was bound to a Master Carpenter for four years at William Porter's yard in Malvern Wells. There were no National hives then. WBC (William Boughton Carr) hives were the ones to make. Near the carpenter's shop, there was a small piece of unused ground and I started there. Hives were put in place on concrete slabs. It was the done thing in those days to house bees in one standard brood box on foundation frames.

A friend of mine, a Mr. Rodway, gave me two swarms early in July which he had run in for me. The bees wintered well. I was seventeen years old at the time. I used to sit watching these marvellous honey bees for hours, coming and going. I know now that these insects had developed over millions of years of climatic conditions and had adapted themselves to the British climate. Mr. Rodway, who had given me the bees and whose father, by the way, still kept bees in skeps, said that if I wanted he would come round one

evening and show me inside the hive. One fine day, a large swarm was hung near a hive. It was easy to shake it into a skep so I hived it separately. This was the first swarm I had ever seen. I told my friend what had happened. So, true to his word, he came to look at them. He examined the stock that had swarmed and found queen cells. He cut out every cell except one and, with all respect to him, I now know that, in doing so, he wrecked the colony. Neither he nor I knew it at the time.

I would like to explain to the reader what the bees would do and how they would react when all the cells but one were missing. After three or four hours, they would be seen running about on the alighting board. These bees would be missing something—not the swarm but the cells. These cells had been carefully made and so arranged that virgin queens would hatch out at various intervals of day or two.

Now, should the colony cast a second swarm or even if, after that, a third swarm (known as a bunt) should issue, the bees would remain in good heart for a new queen would be born—a good one, as nature intended—this is why the natural cells are staggered. They will do this if they are housed in a small cavity or even on British Standard frames that are half blocked with pollen etc.

However if the colony has been interfered with as mine was and all the cells but one cut out and a cast issues then the colony is in dire trouble. The bees know it and build panic queen cells. I have never known them not to do this. Somehow or other, they produce a queen—a poor one which will head the colony. Next spring you will have a poor lot. Leave them strictly alone

bar feeding if necessary and you will find that, by the end of July, they will have superseded themselves so correcting the bee-keeper's blunder. One could requeen it but this is not a job for a beginner as I was then. If there is anything you don't understand about the behaviour of our native bee, please write to me and I will reply.

George Davis, 24a, Lower Wyche Road, Malvern, Worcs. WR14 4ET

Last summer George Davis staged an experiment when he placed a transparent cake cover over a late swarm of bees in one of his hives and was able to watch as the bees built up comb honey. When the bees had finished he was able to remove the cake cover and the honeycomb complete.

Combs in cake tin

BEES CAUGHT IN THE NET

by John Richards

THEY DON'T make it easy, the journalists, banging on about "The Superhighway" without really telling you what it's for and whether it's any good. Let's eavesdrop on a group of internet beekeepers and see what really goes on. You'll meet some users and hear what they have to say. Most of them will be North Americans but, watch out, Europe is catching up fast.

A beekeeper from Alberta takes up our theme, "Folk ask me what good the net is. I tell 'em 'most any question can be answered in minutes if you go to the right place". He's right! There are thousands of net groups catering for all interests and, of course, beekeepers are right there! For the price of a local call you can "talk" endlessly to enthusiasts worldwide. What you "say" appears on screens across the globe; next time you go on line there'll be answers on your screen from all over the world.

Always, there are beginners' questions and—here's the good bit—they always get an answer from old hands. "How long can I keep the hive open and how much smoke should I use?", "How do I stop mould forming in buckets?", "Which way should bottom boards be turned?". Sting-talk is popular. Paulette from LA and Canadian Cindy ask "How often do beekeepers get stung" and "Anyone got a remedy for bee-stings?". "Actually, Cindy," comforts a Californian, "all you can do is grin and bear it!" Robert from Missouri shares his experience: "Provided you don't squash bees and get their kill instinct going or work them late or before a storm you should be all right unless you have hell swarms." A colleague advises, "Have gloves but don't use them. They make you clumsy and carry sting odour. If bees are touchy, try another day. Stings are probably beneficial. Many elderly beekeepers still work like youngsters. In winter, I get backache. I get a couple stings, and the stiffness goes away for a while."

Talk works round to the merits of honey. "Is honey a cure for baldness?" "Nah!", comes the reply, "Jus' makes your hair sticky!" Another thinks: "If honey causes any change, it's an increase in sex drive. My grandad ate lots of honey and grandma said "He went out with a smile". This triggered a memory: "My parents received government surplus honey. The distribution stopped as abruptly as it started. Now I know why! I guess I'll increase my intake."

Look out, here's Liz claiming you can pet bees, if only they will hold still. A rueful Englishman muses, "When I was five, I spent happy days catching bees. Now in my second childhood I'm playing with bees again!"

Townies wonder about types of bee. "I don't care if they're not very productive, as long as they don't attack the kid down the street." says a South Carolinian. Reassurance comes from Virginia: "It takes a little PR with the neighbours but once they have some honey, things go very smoothly." A Texan recommends the Buckfast bee and a neighbour, trying to find the US caretaker for the New World Carniolan breed calls out "Are you there, Susan?" Sure enough, she is and responds immediately from Ohio: "I've been working with them for over 12 years. They are productive, gentle, and good at overwintering."

Someone reports: "Killer bees were first spotted near the Rio Grande-Texas border in 1990. They're not more venomous than other honey bees", he says, "just more aggressive." Thomas from Hawaii adds: "Killer bees are descended from bees native to southern Africa which were brought to Brazil in 1956. Some absconded. Some descendants are the result of matings between the Africans and local European populations." A call comes in from a fire department, "Killers are so difficult to tell apart that we never mess with them. We just hit any urban swarm with foam and kill 'em."

Here's someone wondering whether his colony has become queenless and whether to introduce a young, mated queen. "Don't panic!" he's told, "Just wait. A new queen should be laying within 21 days. Never put a new queen into a queenless colony. Test first. If they draw emergency queen cells, they really are queenless. If the colony is calm, they have a queen. Multiple empty queen cells suggest that you have lost secondary swarms as well as the prime swarm. The queen you probably have will have emerged after the date you expect because the first queens that emerged went with the secondary swarm(s). You need to make an 'artificial swarm' so the bees do what you want, not what they want."

Girt from Sweden wants to thank everyone for all the information which he will pass on to his beekeeping society.

Talking of swarms, Martin complains "In two years, I've removed three swarms from over a dormer and they are back again!" Wayne from NASA comments laconically, "If life gives you a bowl of lemons, make lemonade!"

There are many opinions on smoke. "Dried artemisia calms bees and is less harmful than wood," says one but another cautions, "Artemisia erases bees' memory. It was used in alfalfa fields to make them relearn the territory and work the alfalfa." Talking of alfalfa, "Alfalfa pellets are useful if you work bees all day. They glow like charcoal and produce cool white smoke with little tar." Pakalolo is the favourite fuel in Hawaii! Another recommends, "Well weathered sacking. To make the smoke less unpleasant, add some pine needles. To make it easy to start and prevent it from burning away, leave the ash." Fabio from Italy replies immediately, "I agree: good smell, easy lighting and cheap." But there's also a warning, "Pine needles leave a tarry build-up especially under a slow burn." Another contributor says, "I use rotting cedar bark and add other vegetable matter for a more pleasant smell." Phoebe reckons, "Bailing twine produces nice white smoke with no residue." but someone warns her that, "Twine is often treated with fungicide." Someone speaks in favour of sugar spray instead of smoke but we hear that there is a snag. "It only works well during a good nectar flow. The instant the flow finishes, it induces incredible robbing."

These are mostly lighthearted snips from fuller discussions. There's lots of enthusiasm and friendliness, plenty of weighty material, a lot of detailed argument and many other beekeeping sites to explore! Enjoy!

From Canada, Anthony tells us:
When I was a carpenter, I got slivers...
When I was an electrician, I got shocked...
When I kept bees, I got stung...

I promise, if you start internetting, you'll get enmeshed!

"NOT A LOT OF PEOPLE KNOW THAT"

by Brian Dennis

"There was a gentleman here yesterday," he said - "a stout gentleman, by the name of Topsawyer... he came in here... ordered a glass of this ale - would order it - I told him not - drank it, and fell dead. It was too old for him. It oughtn't to be drawn, that's the fact."

Charles Dickens 'The Waiter'.

I AM LIKE Autolycus in Shakespeare's The Winter's Tale "a snapper up of unconsidered trifles" or what my more critical friends call "a fund of useless knowledge". In particular, I collect snippets of beekeeping lore as avidly as any philatelist! Did you know, for example, that clergymen live longer than most? It is too soon to know whether 'clergywomen' will equally benefit from their calling. But beekeepers live longer than clergymen. It follows, therefore, that beekeeping clergymen must have an enviable life-span. As Michael Caine says: "Not a lot of people know that"—and like most generalisations, such a conclusion may contain, at least a grain of truth.

Nowadays, in this health conscious, health obsessed age, much of what brings pleasure and relaxation in our increasingly stressful lifestyles is bad for us. The fact that experts disagree and their collective wisdom keeps changing doesn't help. I have known women as thin as lathes on slimming diets, convinced they were overweight and overweight men who, after years of being sedentary and eating, drinking and smoking to excess, took up jogging—and dropped dead from heart attacks. So much for mens sana in corpore sano—a more hedonistic lifestyle is my aim—my ambition is to be shot by an irate husband at the age of 90! After all, an expert is only 'x' the unknown quantity and 'spurt' a drip under pressure...

This set me thinking about earlier times when life was hard. I have a snippet that refers to the author's great-grandfather who kept bees, made his own mead, and smoked an ounce of thick twist a day since

The Smoker

the age of 13. This great ancestor was against things—mainly Temperance Societies! It was only after he had outlived all his opponents that he decided, at the age of 107, that he had had enough and quietly departed. Incidentally, mentioning Temperance Societies brings to mind the origin of the word teetotal. A proponent of total abstinence from intoxicants had a stutter. When he addressed meetings, he referred to t-t-t-total abstinence. Hence, teetotal! However, did the drinking of mead contribute to great-grandfather's longevity?

As I wrote in my previous article, the claims made for mead are almost as many and widespread as those for honey. But honey defies scientific analysis—although many components have been identified, including essential minerals and vitamins, there is nothing in such abounding quantity to explain the great benefit that honey and mead seem to promote. There is almost a perverse satisfaction in having

a mystery trapped in a bottle of excellent mead. The only scientific evidence that is often cited is that of the Colorado bacteriologist who transferred typhoid bacteria into honey to see if it would act as a suitable medium for breeding further quantities of the bacteria. To his surprise, within 48 hours the bacteria were dead. In a later experiment, some honey was impregnated with the combined bacteria of typhoid fever, paratyphoid, enteritis, dysentery, bronchopneumonia, septicaemia and peritonitis. Within five days not a single bacterium remained alive.

Until about 1750, honey was the principal sweetening agent. It was not until the Industrial Revolution, when machinery was invented to process sugar cane, that sugar ceased to be a luxury enjoyed only by the rich. Honey was abandoned and mead making, already undermined by hopped ale and the importation of cheap spirits and wines, went into decline.

"Rub honey on head to cure baldness..."

It works !!

BPD

69

In more recent times, it has been suggested that many ailments can be attributed to, or worsened by, a large intake of white sugar. Much 'evidence' is cited in books promoting the health-giving and curative properties of honey, but most appears to be folklore rather than scientific research. But Professor Jung has pointed out that if one finds a belief in several unconnected cultures at different times in history, one should be wary of dismissing it out of hand. Some claims can be easily verified. Massaging the scalp with honey is said to cause the shiniest pate to sprout hair. Anyone like to check this one out? However, beekeepers—even the less hirsute—appear to be a healthy lot. A curious fact arising from all this is that quite small quantities of honey or mead are needed to produce large health benefits. A couple of glasses of mead a day will assist one's health—once one has been drinking thus for about a year. I drink mead for my health—it is medicinal!

I hope you had a go at making some mead following my article in last year's Beekeepers' Annual. While that is maturing, why not try some quicker honey brews? If you already make beer you will know how to go about it. If you don't, talk to someone who does—the process is simple enough but home-made beers are conditioned (made gassy) by adding a small amount of sugar to each bottle. If beer is bottled too soon or too much 'priming' sugar is added, the result can be like Vesuvius or a fire extinguisher on opening the bottle. It can also result in burst bottles which, of course, is highly dangerous if you are holding it at the time. Do use beer bottles designed to withstand pressure.

HONEY BEER

Ingredients:

4 oz. honey
1 lb. sugar
1 oz. ground ginger
2 fl.oz. lime juice
Juice of 3 lemons
1 gal. water
All purpose wine yeast

Method:
Boil 4 pints of water with the ginger for half an hour. Pour onto the honey, sugar, lime & lemon juices in a suitable container. Stir to dissolve. Add 4 pints cold water and when cool, add the yeast. Ferment, closely covered, until fermentation has finished and then bottle and prime.

The Celtic name for Britain was 'The Honey Isle of Beli'. The Celts sometimes added ingredients to their mead to impart magical properties. For example, they fermented a mixture of honey and the juice of either the hazel or birch tree which they believed endowed them with superhuman strength—as a result of which they probably went about pillaging their neighbours! But one of the good things about mead is that it is not a depressive, unlike many other drinks, and they undoubtedly pillaged with great joy and exuberance! The following recipe should help you get the best out of life but please, no pillaging...

CELTIC ALE

Ingredients:

7 pints birch sap
3 lb. honey
$^1/_2$ oz. malic acid
$^1/_4$ oz tartaric acid
(or $^3/_4$ oz. citric acid - 2 level teaspoons)
Tannin (or cup of strong black tea)
White wine yeast or All Purpose yeast
Yeast nutrient

Method:
Obtain the birch sap. In early March, drill tree with diameter over 8" - 1cm hole, 2cm deep (just beyond bark), about 30cm above ground. Insert a length of plastic tubing—plug neck of collecting jar with cotton wool. It may take about 1 week to obtain sufficient sap—plug the hole when enough sap has been obtained or the tree may bleed to death. Add the honey and other ingredients (except the yeast), stir well to dissolve the honey and add 2 Campden tablets. 24 hours later add the yeast and ferment to dryness. Rack and top up with water. Mature 1–2 years. Drink in small glasses if you don't want trolls dancing on your head in the morning—this is powerful stuff! This is more like a wine and does not require priming.

Mead was drunk on great occasions such as a wedding feast but for everyday use a low-powered mead-ale was made. When ale from honey and ale from malt were both available, that made from honey was considered superior. It was only the increasing cheapness of malt-ale which caused honey-ale to go out of general production.

ANGLO-SAXON MEAD-ALE

Ingredients:
1 $^1/_2$ lb. honey
$^1/_2$ oz. hops
Juice of 3 lemons
Beer yeast
1 gal. water

Method:
Bring the honey to the boil and skim. Add hops and simmer for 15 minutes. Strain. Add lemon juice and ferment. When finished, bottle and prime (drinkable after about 8 weeks).

References to mead, sex and long-life are many and world-wide and I quoted some in my previous article. The Scots have a saying that mead drinkers have as much strength as meat eaters. In Scotland centuries ago, a wasting disease was treated by a concoction known as Athol Brose, which contained heather honey and Scotch whisky, taken little and often.

ATHOL BROSE

Ingredients:

3 heaped dessertspoons oats (not "instant")
8 oz. cream
8 oz. whisky
3 dessertspoons sherry
2 dessertspoons clear honey

Method:

Cover oats with water and soak overnight. Strain through muslin. To the liquid ('brae') add the cream, whisky, sherry and honey, stirring after each addition.
Bottle and refrigerate for 5 days. Remove an hour or so before use, let it thaw a little, pour into a bowl, stir and ladle into glasses. This does not keep indefinitely.

Having made your mead (or honey-based drink), you should, perhaps, follow the ancient law of Ireland: "There are three things in the court which must be communicated to the king before they are made known to any other person: first, every sentence of the judge; second, every new song; and third, every cask of mead."

In conclusion, the sad and tragic story of the Norse king Fjolne. A great feast was prepared for him by a prince called Frode who had built a great timber mead tank. Fjolne dined well and had to be carried to his bed in one of the lofts. "In the middle of the night," the story relates, "he went out into the gallery to seek a certain place." On his way back he stumbled into the wrong loft and fell into the mead tank and was drowned. It illustrates the fact that you can drink mead almost indefinitely provided you remain seated. As an old Polish proverb puts it: "Mead makes you drunk from the waist down." Wacht heil.

CARRYING ON THE WORK

by Ron Coles

 NOW DON'T get stung", those were the first words of advice given to the young Peter Donovan when, aged 13 years, he appeared at the elbow of the man who was to become his mentor, Brother Adam. Peter had been evacuated to Devon when the war had threatened to make London a dangerous place to live and he found himself billeted in an attic room over-looking Buckfast Abbey. Fascinated by the sight of the monk working the beehives in the apiary in the grounds of the Abbey below his window, he had plucked up the courage to go and get a closer look.

Of course, he found it difficult to take that advice, particularly as there were no bee suits available then for protection but it did not seem to put him off. He became a regular visitor, much to the annoyance of his guardians who objected to him coming home sticky and smelling of smoke. By the time he was 14 and able to leave school, he was working with brother Adam officially, earning the princely sum of £1 per week, This idyllic lifestyle came to an abrupt end, however, when Peter's call-up papers arrived. He soon found himself in Malaysia where he became popular chasing the local honeybee and producing mead for sale to the Officers' Mess, using Brother Adam's recipe.

Peter Donovan

After the war Peter took a job that paid better than Buckfast Abbey could afford, back in his native Kent. He kept in touch with Brother Adam and set up his own apiary using Buckfast stock. After some years he saw that the Abbey was, once again, looking for a beekeeper and this time he managed to negotiate a decent wage which enabled him to take the job. The hours were long and his boss was a tough taskmaster, driving his staff and himself hard, but Peter enjoyed it and he learnt a lot. That was 25 years ago and he is still there, carrying on the work of Brother Adam, although he is now past normal retirement age, himself. Not that he looks it. Try as one might, however, Peter will not be led into putting his fitness down to eating honey or royal jelly, he says it is the same as with the bees, it is all down to genetic material! Not much hope for the rest of us there but then perhaps sensing

73

my disappointment he added that he has enjoyed good health all the time he has been at Buckfast and is the only member of the Abbey staff who can claim not to have missed a day's work through illness for the past quarter of a century.

It is perhaps as well, there is much work to be done. There are 328 honey production hives and in the summer 500 nuclei to be looked after by Peter and his two helpers. The hives are set up in groups of four facing all points of the compass because Brother Adam noticed that when the hives were placed in straight rows facing south, the end ones did well but the middle ones suffered. In fours the problem is eliminated and the North facing hives seemed to do just as well as the rest even though they are slower to start in the mornings, waiting for the sun to warm them up. This arrangement has the added advantage that the adjacent hive can be used as a table when working on the neighbouring one and you never walk in front of the hives on which you are working. Each hive has an alighting board to help the bees get into the hive quickly, when they are loaded up and heavy with nectar and pollen, safe from predators like wasps. The hives are maintained on a four year cycle, stripped, filled and painted with Cuprinol inside and an external emulsion outside, As a result, some of the hives predate Peter's time at Buckfast They are still in good repair after 60 years of constant use.

Every week the colonies are inspected and Peter reckons that they now have it down to a fine art, taking four to six minutes per hive including the removal of unwanted queen cells. Peter tells his two assistants, both young monks, that the hive is like a book, the combs are the pages and it is possible to read on them all the information you need to understand what is happening inside. Furthermore, no two hives are the same even though the queens may be sisters.

The breeding programme is the same as the one Brother Adam established. Quietness, uniformity, low propolis production, and the ability to winter well are key factors in Brother Adam's list of 25 desirable qualities. Bees that dart about on the comb or fly at you, produce wet cappings or build bridge comb will not be chosen. Peter assesses each queen every three weeks and compiles records (computerised today) in order to select the queens for breeding purposes. Full

Peter in the Buckfast Apiary

information is stored and each queen ends up with a pedigree showing the line, just like the kind you would expect to see if you bought a pedigree dog. In fact there is more information if anything. Nuclei are weighed every March to see how much food they

Peter with Brother Adam

Four hives together

have stored for the winter, any light ones will not be used for breeding.

Good ones will be carefully placed in the apiary to see what other credits they can notch up during the course of the season. They are already destined to be chosen if they can do well in the other key aspects. When they are selecting for breeding the queens chosen will be six sisters from the most highly rated stock according to the criteria already outlined.

All of this work is done purely in the interest of quality beekeeping at Buckfast Abbey. In fact, if you want to buy pure Buckfast strain breeding queens you will probably have to get them from one of the licensed suppliers abroad, These dealers import genetic material from

Buckfast to keep the strain pure but after several years are mostly self sufficient. The natural queens which have randomly mated in the apiary with unidentifiable drones are available for purchase but as Peter says there are not too many or the Brothers would get it in the neck for not doing their job properly. Anyone buying such a queen must undertake not to breed from her and sell on as Buckfast queens.

Although beekeepers always talk about the quality of their queens, the importance of the drones must not be underestimated. Peter insists that 72% of the genetic influence is produced by the drone side of the family. So whatever the strain of the queen he always uses dependable drone lines of Buckfast origin. This guarantees the production of quiet bees.

For this reason there is a market in Buckfast drone semen, Peter had recently despatched five capillary tubes of semen to America. The capillary tube used for storage and travel is also the syringe used for insemination. It had taken a hundred drones to fill each phial and the contents would inseminate eight queens. The semen is collected in about three to four hours, it goes in the afternoon post to the States, taking about five days from collection to use but in fact, Peter believes that the semen remains in good condition for up to ten days.

The hives in the Buckfast Apiary

Whether they are given a helping hand or carry on as nature intended with just a little management input from Peter Donovan, there is no doubt the Buckfast queens will continue to raise good natured colonies producing large quantities of honey. It is also interesting to think that in a number of distant corners of the world miles away from Devon, they too are producing pure strains of the Buckfast bee. Maintaining the line of the bees is important but with Peter's retirement only a couple of years away, so is maintaining the line of beekeepers.

The collection of the semen is a tricky operation conducted under a microscope. Artificial insemination in the hands of an expert is a good way of holding the line but as Brother Adam says you can never be absolutely certain of the genetic material you are using and nothing can replace a perfect natural mating.

Already the Abbot is having to think about the succession, I cannot think of anyone better than Peter Donovan to pass on his enthusiasm and the complex skills of the craft. I feel sure that one of his assistants will pick up the baton and follow Peter in carrying on the work of Brother Adam.

CROSSWORD

Compiled by Byron

ACROSS

1 (4) A simple starter.
3 (8) Soldier in a shock troop.
9 (5) Language used by bees.
10 (4,3) Sting in the tail.
11 (3) A drowned valley.
13 (5,4) Protection for fireman - and for beekeeper?
14 (2,4) Time for bees to be back home.
16 (see 24 down)
18 (4,5) It's to your advantage to be one of these.
20 (3) To be told on sight.
22 (7) Half a doublet.
23 (5) My mate, the Chancellor.
25 (8) Domesticated caterpillar.
26 (4) Insects without number have___and gone.

DOWN

1 (5) A down clue is appropriate for this bird.
2 (3) Apian compass.
4 (2,4) Fiddler and tiler could meet here.
5 (7) Most selfish.
6 (2,7) No chance of divorce! Coffee? Do you mind instant?
7 (2,5) Where clothes are never seen.
8 (4) In short, a fine queen.
12 (9) Ventral.
14 (7) Irish profanity.
15 (7) A bird to make you gulp.
17 (6) Drink of the gods.
19 (4) Some do it with doughnuts.
21 (5) Party with yearning makes leader of the gang.
24 and 16 across (3,6) Altogether nourishing; separately hurtful.

BILL GETS ELECTRIFIED (CHEAPLY)
by Bill Clark

JANET'S TWO FRAMER

 NE HOT SUMMER evening back in the mid seventies, I was delivering a Newsletter to a CBKA member, and upon my arrival Janet gratefully stepped back from her little, 1930s, two frame extractor. 'Ah Bill! What an opportune moment,' she exclaimed, 'I was just going to get a cooling drink. Be a dear, just spin these out for me whilst I get it.' And with that she departed, only pausing to call out towards a further out-building, 'I am getting a drink Harry, Bill Clark's here.' Dutifully I spun the little extractor. It did the job well, but because of the small diameter of swing, it had to be fairly whizzed round, and, coupled with the fact that the drive cog on the handle had only twice the teeth of the spindle, I found my hand was going round at quite a rate too. No wonder Janet looked ready for a drink!

She returned with the drinks, 'I bet you are glad you don't have to fiddle with a little thing like this,' she said, aiming a kick at the extractor, 'but it is just not worth a lot of expense when I only have two hives.' I deemed it best to commiserate, and not point out that although I had a four frame hand extractor, I had to spin out the honey from nine hives! Then Harry came in with a new

No.1 – Is a copy, with a similar drill, of the little two frame extractor.

looking electric drill in his hand. 'Here you are Bill, what do you think to this little beauty then?' he said, and passed it over. The drill was the chunky little 'Bridges', but what Harry particularly wanted me to notice, was his own adaption of an electronic speed reducer fitted in the cable; through this he could dial up from something like twenty revs per minute, to a full three thousand or so. Harry was a boating enthusiast, and spent hundreds of happy hours drilling holes. No, I don't understand it either, but it is apparently a major part of a boat builders life! 'Now,' said Harry proudly, 'with the right tool tip I can drill anything you care to mention!' I swung round to Janet, drill in hand, 'I am your fairy godmother, you shall have your electric extractor,' I said. Poor Harry's face blanched as I explained—it was only a matter of moments to take off the worn cogs, then tighten the chuck of the drill onto the centre spindle. As luck would have it, the hole where the handle had been, was in just the right spot to take an upright bolt to be the stop for the drill handle. The happy look on Janet's face said it all, as minutes later the next two frames were spinning effortlessly around. However, when I noticed Harry's anguished look as the drill handle chattered against the stop, I did think it best to make one small change, and placed a rubber band around the drill handle and stop. With Harry now looking happier I deemed it time to wend my way homeward.

BILL'S FOUR FRAMER MOTOR DRIVE

A day or two later sweating away at my own honey crop, I found my thoughts turning longingly to Harry's electronic speed reducer, and by the end of the evening I was ready to kill for it!

I got in touch with the supplier, Proops Manufacturing, Unit I0-11, Wharf Lane, Burton on Trent, Staffs., and gave them the details of my most suitable electric motors.

No.2 – Is of the parts that I used to convert a spin drier.

(Readers of my other DIY articles will already have guessed, I have some half dozen, rescued off various washers, spinners and pumps, languishing in my shed) Two of the motors could be used with the reducer, and I sent off my £6.50. This was the only money spent. Yes I know, I must be slipping! But it was obvious that Harry was not likely to give up boat building in the near future, and I am not really of a murderous bent, well, only sometimes. (You will have to pay £23.79 to include VAT, and it is now a square shape)

It all went together rather well, and as you can see from the photo, the parts list is quite short.

- One suitable pulley wheel, (size is immaterial, but as large as possible without making it difficult to put frames in and out.)
- One aluminium saucepan, or similar redundant item, of about 10mm larger diameter than the pulley wheel.
- One fan belt that will stand proud when pulled tight around the pulley wheel.
- One small electric motor. (Mine is from a spin drier).
- One bush, about 12-15mm diameter. (Mine was the fan off another motor, with the blades ground off)
- Some plastic beading or a length of small bore tube. (Mine came from the edge of fridge shelving)
- A collar for joining plastic drain pipe.
- A piece of expanded aluminium mesh.
- 150mm x 25mm x 2-3mm thick piece of metal.
- And of course the speed reducer.

First dismantle the cogs and handle from the extractor, and put them safely away, in case of break-downs. (My pulley wheel needed the centre bored to a larger size, and very conveniently had a small grub

screw in place, ready to tighten onto the spindle.) Choose a suitable V belt, now cut it to a close fit around the wheel. (Cuts best made at an angle). Remove and put a few spots of a good all-purpose glue around the wheel, and some puncture outfit glue on the cut ends of the belt. After the required curing time, put it back around the wheel, with a string tourniquet whilst it sets. Slide the wheel into place. (I see no reason why a small rubber tyred buggy wheel shouldn't also serve the purpose.)

My spin drier motor has a side pivoting mounting, and a spring that kept the motor pulling away from the wheel, and tight in the belt. It was very simple to reverse this, as can be seen in the photo, and the motor now bears against the pulley. The beauty of this is twofold, it allows for the 12 to 15mm diameter

No.3 – Is the spin drier motor conversion ready for use.

bush to be fitted to the end of the motor spindle and used as the drive, so giving more revs to the motor and the cooling fan, also it can slip on take up, like a clutch, giving no snatch. If there is excessive spin, just tighten the spring towards the pulley. A hole drilled in the cross bar allows the motor to be bolted in place. Any other kind of motor will need more juggling, but it should be no big problem to fix, say, a door hinge in the right place, and a pulling spring across an angle, rather than my twist one.

Now measure the depth that is needed for the guard, that is, from the bar that it will be fixed to, to a point about 30mm or so above the top edge of the wheel. Cut the saucepan at that depth, and also a 25mm hole in the centre. Take off the wheel and put the guard in place, at the point where the motor drive spindle will come against the wheel—a small segment has to be cut out to allow contact. All that remains is to replace the wheel, position the guard so that it doesn't rub, and

drill for two self tapping screws or bolts through the bar. You can now tighten up the wheel. You could fit the saucepan lid back on, with a suitable piece cut out for the drive and two or three self tapping screws around the edge, I just put plastic beading round the sharp edges.

Because my motor had an outside fan, I had to devise a guard. The plastic drain collar happened to be to hand, and the right size! I first cut off one outer band from one end, and pulled out the rubber rings from both ends, then cut a circle out of the expanded mesh using a rubber ring as a template, hammered it into a dish shape, slid it into the place whence the ring came from at the undamaged end, stood the collar up, and gently flattened out the mesh with the end of the hammer handle to fit it snugly in place. Warming the collar with my blow lamp, I pushed it onto the non circular motor. After it had cooled to shape it only remained to mark the place for the cable to come through,

No.4 – Is yours truly on the door step with the cheapo drill stand.

take it off, drill, fit a cable grommet, thread through the cable, put on a plug, replace and screw in the holding screws. There are many types of motor, and this can well be one chore that you will not have to deal with.

Next take the piece of metal, and bend at right angles (some 30mm) from one end, so that the short length covers the place where the bolt fixes the cross bar to the edge of the extractor, and the long side sits snugly down the outside of the tank; this bolt is now going to also fix down the angle piece, make sure that you drill the appropriate size hole in place to allow this. When all is OK fix the speed reducer to the bar, making sure that it doesn't stand too proud so as to foul a wing nut if one is used. You should now be at that happy stage of being able to switch on. Mine has surpassed all expectations, not a single problem in fifteen years.

A DRILL CONVERSION FOR THE NINETIES

During the time that I mulled over whether there would be enough interest in my motor conversion, I was very mindful that most people have got to find at least an understanding electrician, if not a complete engineer. I thought back to Harry's drill of fifteen years ago, and how simple the job proved to be. My eyes lit on my present one reversing, hammer, torque control and built in electronic speed control. The 550 watt power must be about double the little old 'Bridges'. Most folk now have something similar, and probably even more powerful. Surely this would do the job, and without all the engineering fuss?

You can of course use an ordinary drill with the electronic 'proops' speed control. And so dear readers, in your interests only, I stripped down my beloved extractor, fitted on my drill, and with four full frames inside—it indeed made no fuss at all in the spinning,

and covered all the speeds needed with ease, but was so jerky in action. And I worried in case the hard start up should burn out hundreds of drills world wide. Mind you the drill manufacturers might give me a pension!! I then realised I was back to the problem that I encountered with the motor conversion. I needed a swivel mount, and a slip clutch. A hasty search of the workshop, and even a visit to to my favourite recycling centre, didn't turn up anything that could be available to all and sundry; so there was nothing for it! I should have to visit Texas Home Care. 'Oh Bill wash your mouth out with soap and water.' Well it had to be done. I do not stint myself where the readers of this Annual are concerned. Anyway skipping over this painful part quickly, suffice it to say I found exactly what I needed—a cheap drill stand, one that I wouldn't be seen to carry home normally—costing under £10.

No.5 – The electronic drill conversion, using the drill stand.

Now down to business, all we need is the all purpose drill cradle, and its attached sliding stem. You can give the pillar etc. to the kids to play with.

- The peddle crank arm from a cycle, not the modern square hole one.
- One baked bean tin.
- One spiral spring, similar to the ones under some cycle saddles.
- One piece of metal 150mm x 12mm or so x 5mm or so thick.
- One aluminium saucepan.
- One pulley or buggy wheel to fit the extractor spindle.

Make up and fit the wheel to extractor spindle, as in the previous conversion. Now fix the drill into the cradle, and offer the sliding stem into the large end of the pedal arm. It is a sloppy fit? This does not matter for the moment. Next stand the whole thing

No.6 – Same as five, but showing me fitting the tension spring.

on the extractor cross bar. Lay the pedal arm to allow the stem to slide down through and past the cross bar, (you may need a helper at this juncture) swing the drill to touch the wheel, then mark the best spot to drill through both the arm and the cross bar, drill hole and bolt on. Now slide up the drill and stem until the smooth part of the chuck, (with the holes for the chuck key) is bearing against the wheel; make sure the teeth are clear of the wheel edge, for this is to be the drive against the extractor wheel. Now measure the gap between the base of the drill clamp, where the stem fits in, and the surface of the pedal arm, where the stem slides in there. You now need a piece of metal tube or a stack of washers to fit this gap exactly. When all is satisfactory, mark where the surplus piece of pedal arm needs cutting off and disassemble. Cut off the surplus piece of peddle arm, slide the tube or washer spacers onto the stem, cut a piece out of the bean tin and fit it spirally round the end of the stem, and tap it into the piece of arm. You may screw this up once or twice, but try again, bean tins are relatively cheap!! The stem should now be a nice fit in the arm. You can now leave it in this condition, or having once again offered it up to make sure all is well at the point of contact on the wheel, get it welded, for the drill cradle can swivel on the stem anyway. At this point follow the procedure for fitting the saucepan guard in the previous conversion. Lastly make a small crank in the piece of metal strip, drill a hole to coincide with the cross bar fixing hole of the extractor and another at the top to fix the spring across keeping the drill against the wheel. (It is possible to fit a twisting spring around the stem to make a neater job)

The whole job took me one and a half hours, and it works like a dream, cogs, both cast iron and plastic, have had it from now on where I am concerned. You now have three different ways of saving your energy, and my wife may even get her spin dryer motor back!!

HAVING A FIELD DAY

by Paul Smith

LINCOLNSHIRE B.K.A. WRAGBY FIELD DAY A POTTED HISTORY

WE THINK OUR Field Day here at Wragby is one of the oldest Open Days still running in the country. The first event was organised by Miss Nancy Ironside at the local Vicarage. (Miss Ironside, incidentally, organised the 14th Apimondia Congress in Leamington Spa in 1951.) It was not until 1961, however, that the event really took off. The venue was the local Town Hall (built by the building firm of E.H. Thorne Ltd.), with a series of lecturers, one demonstrator, and the famous strawberries and cream teas. In that year we even arranged a visit to the local injection moulded plastics factory.

The first speaker we ever had at the event was Mr. T.H. Hallam of Staffordshire, with Mr. Don Sumpter of Northamptonshire manipulating the bees.

The format for the event in the early days was firmly set. It was always on the second Saturday in July, starting at 2pm with an official opening (one year it was the Euro-rebel Tory MP Sir Richard Boddy).

This was followed by two one hour lectures and a bee handling demonstration: the latter cleared the Hall so it could be prepared for a 5pm tea. At 6pm we had the Brains Trust, made up of lecturers and the demonstrator of the day.

In the early days it was very much a family day out with children being able to play in the adjoining park and playground. The year that Bob Couston was our guest speaker he did his famous magic show for the children and brought the house down.

We must not tempt fate—BUT it has never rained during our Field Day.

It did rain early in the morning in 1961. Freda Adams had volunteered to pick the strawberries at that time and got VERY wet in the process. Since then the strawberry picking has been done on the Friday afternoon/evening, and in the last ten years or so by the ladies in our sewing room.

Our list of distinguished speakers will be recognised by many of you. Alas several are no longer with us: Ken Stevens, Capt. Tredwell, A.S.C. Deans, Bob Couston, Robert Creighton, F. Padmore, George Smith, Cecil Tonsley, Don Sumpter, Eddie Eade, Ted Hooper, George Hawthorn, George Knights, John Atkinson, Commander Dixon, Robert Pickard, Robert Paxton, Peter Beckley, Norman Rice, Len Heath, Edward Crimmins, Bob Hammond, Medwin Bew, Paul Wix, Clive de Bruyn, David Charles, Andrew Matheson, Robert Park, Vince Cook, Michael

1995, Some of the 'Early Birds'

Solley, Matthew Allan, John Cossburn, Gerry Collins, and many more.

Over the last few years the event has split into two distinct parts. The Open/Field Day beginning at 11am and our massive sale of equipment, at knock down prices, which starts at 9am. For several years now some beekeepers have been here at 5am to ensure they are through the doors first! When the doors open it's every man or woman for him/herself. Last year we estimated that over the day we had 500 people attending, (admittedly some had left by 9.15am!) and unless you know better, we think this is the second largest single day event in the UK after Stoneleigh.

We try to vary it year by year, and latterly we have added some exceptional demonstrations to the programme (cookery and bee handling can sometimes be tedious), Wax Flowers by Liz Duffin, Candlemaking by Margaret Angus, Skep Making by Paul Hand and Larvae Grafting by Ben Gillman.

Our factory is working all day, so as well as learning about beekeeping, watching beekeeping, buying beekeeping equipment, you can actually see it being made—who else does that?

1976, Jean Purcell demonstrating, Bob Hammond kneeling right.
This must be the Seventies, look at those flares

JUMP ON YOUR CAMEL

by John Kinross

AN ENTIRELY NEW and painful experience" is how Julian Johnston describes his experience, not keeping African bees, but riding a camel in Southern Arabia. His book "*A Nomad amongst the Bees*" has just appeared from NBB (**£5.50**) and makes very entertaining reading.

I know what he means and also some of the area he describes in his book having spent some of my National Service in Aden. However we didn't have to rely on camels—only on the odd lift from the CO's chauffeur or Shank's pony. On one occasion we went to see some civilian friends by sailing boat, which was fine on the way out but difficult and dangerous on the way home as the wind was against us.

Aden also boasted the odd lizard usually with a long tail and I have also seen them in

Malta. So it was quite a surprise to see the lizard illustrated in Dennis Anslow's "*A Beemaster's Tale*" (**£3.50** from BBNO).

Dennis is an artist as well as a Staffordshire beekeeper. He says, that Staffordshire is one of the few inland counties that has good wild flowers, plus heather and ivy for bees. And Old Bill his beekeeper friend comes from a long line of beekeepers so that his stories, all illustrated include Mary Queen of Scots who was imprisoned at Tutbury for a time (where our new BBKA Honorary Secretary comes from). I visited the castle there recently to find that one has to be a shepherd to get in as the gateway was full of animals, not camels or lizards but sheep so the mowing bill did not trouble the accountants.

From BIBBA comes another detailed little handbook: "*Bee Breeding & Queen*

Encampment in desert - A Nomad Amongst the bees by Julian Johnstone

Rearing" (**£3.00**), which has a foreword by Adrian Waring and four articles: Native Bees by the late Beo Cooper, Pure and Cross Breeding by Dreher, Bee Breeder and Propagator by Ken Ibbotson and the Stahl Method translated from German by Tibetan Carpet expert P. Denwood. Where are the camels you say? Well there aren't any but one paragraph talks about Genes,even if they aren't the the same variety that came out of Aladdin's lamp,

For lamps and night lights (presumably nitelites in USA?) you must get the new edition of Ron Brown's *"Beeswax"* (BBNO **£9.95**) which has a colour photograph of Rosemary Brown's horse on the front. Ron spent part of his war-time service in Gibraltar and recently I went there to see if I could visit his old office. I must have got close to it as it was underground, but alas I had no candles and most of the tunnels are closed to the public. There appear to be no bees on the promontory—it is not an island although included in the David and Charles Island Book Series—but it does have some wonderful spring flowers like sea lavender, aconites and giant squill. It also has lizards, although, for those who know, the guide book calls them Iberian Wall Lizards, Moorish Gecko and wait for it, a Large Psammodromus. I know what that one is like as I shared a room with an overweight American who snored all night.... The book to get on flowers of the area is Anthony Huxley's "Flowers of the Mediterranean", now believed to be out of print.

Africanized bees are included in the new edition of Roger Morse's *"Rearing Queen Honeybees"* (BBNO or B&K Bks **£10.00**) and there is a paragraph and photograph of Brazilian bees, where Warwick Kerr uses Africanized bees. They do sound fierce and it doesn't help having photographs of unprotected beekeepers manipulating them. However the book is easy to read, has a useful index and also a bibliography that mentions some other useful books on queen rearing including Doolittle, Pellett and Laidlaw. One English book is listed: Snelgrove's *"Queen Rearing"* which is still available today (BBNO **£8.95**) and his famous book *"Swarming"* is about to appear in a 14th edition with an Introduction by Karl Showler, who has been able to throw some light on Mr. Heddon, inventor of the Heddon Method (BBNO **£5.95**).

Returning to the land of the camel, I have found an article by an unknown clergyman in the August 1886 edition of BBJ: "The colour of the native bee is somewhat like that of the inhabitants... A few slabs of stone so adjusted to form a little house suffice for their bees. What is worse than their so called hives is the way in which honey is taken. It is a mere mixture of mould, dead bees, honey and wax. The natives make it with a fermented beverage. It will be a long time before it can be said that of this area as of ancient Palestine," it was a land running with wine, milk and honey.

Maybe he should hop on his camel and come home. Now it is time for the near end-of-article joke. It was told to me by a local policeman.
A famous beekeeper died and his widow went to a seance and managed to contact him. Assuming he was in Heaven, she said:-
"What is it like there Douglas?" (He was a Scot)
"Wonderful I eat all morning and breed all afternoon with a nice long nap."
"I didn't know Heaven was like that Doug."
"I'm not in heaven."
"Well the other place then."
"Don't be silly old girl, you are talking to a rabbit in Bishops Stortford."

The most important book of the year for me is *Pollen Analysis* by Moore, P.D., Webb J.A. and Collinson M.E. (Oxford £24.50) which appeared in a different format a few years ago published by Hodders in a chocolate brown cover and without Collinson. She has really made it a new book with a striking blue and yellow cover and no less than 71 pages of black and white pollen photographs. These are not all to the same scale so it is difficult to work out sizes, but for those of you who like technical details: the dried pollen was examined under a Phillips 501B SEM microscope, the image digitised with a Tracor TN S400 imaging system to a resolution of 512 x 512 x 8 bit (equals 256 grey levels). The image was transferred to an Acorn Archimedes microcomputer for editing, colouring and photography.

Thus you can use an Acorn to see inside an acorn. This is a book that is a bit frightening to the non-scientific but I am told is a must for all those of you who look through microscopes at honey plants. I once tried, using Rex Sawyer's card system, to look at dandelion through a microscope—I worked it out backwards via the cards with lights showing through uniform holes to the fact that the spore was from a dandelion and felt very pleased with myself that the image did look like the picture in the book, but I was unable to do it the other way round, ie. using a flower that was unknown to me.

I do recommend if you buy a microscope that you get a powerful one and not a little job that mists up when you look through the lens. I once was privileged to look through the Electron Microscope at the RAF Tropical Diseases laboratory at Halton, Buckinghamshire close to where the late Fred Jones lived. Unfortunately I wasn't permitted to make up some slides for Fred's library but had I been, I can just imagine the scene when Mid Bucks Beekeepers sat round Fred's table to see his latest slide:

"What's that funny red thing Fred?" asks a member.
"Not sure, but I'll read the words on the slide," says Fred.
"It reads Airman infected with Montezuma's Revenge, Aden 1956," says Fred. Some slight laughter from the back row, possibly from some ex-National Serviceman who just happened to be there at the time and recalls that week in Steamer Point Hospital with Air Marshal Sky-Whyte always keen to get blood samples for his microscopic collection. After all Montezuma's Revenge was a usual and very unpleasant disease which is common today not only to airmen but to all tourists who travel beyond Calais.

Finally we should mention the Nutshell Series. Handy for all squirrels, the NBB Nutshell leaflets by Matthew Allan are with us: 4 titles so far "Manipulations", Starting Out", "Fun with Bees" and "The Beekeeper's Toolbox" at £1.50 each you cannot go wrong. Full of practical advice with a colour cover all they need is a binder and index and some information on future issues. BBKA have been crying out for a cheap beginner's leaflet for years and maybe this is the answer. Strange that we have used a Scottish BKA booklet for this purpose until now, and along comes a Scot with retail experience in Hampshire, who has used it all to good effect. Come on BBKA, add your seal of approval and then we can recommend it to all beekeepers' associations

C.B. DENNIS

by David Little

IT IS WITH great sadness that we have to record the sudden and unexpected passing of Brian Dennis, aged 84, at the Princess Elizabeth Orthopaedic Hospital, Exeter, on 10th August.

Born in Blackpool, his family moved to Germany, then to Newcastle in 1922. Brian joined Kodak in Harrow around 1930, where he had a successful career as an analytical chemist, and it was here that his long association with BBKA started. A member of the Executive in the 50s

and 60s and again from '77-79, he represented BBKA on the MAFF Bee Disease Advisory committee and the British Standards Institute Committee; he was also involved in the reform of the BBKA constitution in the late '50s, and a keen member of the Central Association. His knowledge of Bee diseases made him a valuable and well respected member of the BBKA Research and Diseases Committee which he served from the 50s until this year, being Convener and then Secretary for some 11 years. He was instrumental in the production, by KODAK, of the first Honey Grading Glasses.

He was a founder member of BDI around 1930, secretary for the next 50 years and their delegate to the BBKA ADM for most of this period. His contribution to British Beekeeping, particularly in the field of disease, was recognised in 1984, when he was made an Honorary Member of BBKA.

He retired to Bideford in 1971 where he added the teaching of Silver-smithing and gardening to his other activities; he was renowned in the area for his collection of Orchids.

In 1990 he established the C.B. DENNIS BRITISH BEEKEEPERS RESEARCH TRUST, which uses the interest from its capital to support British research projects likely to benefit beekeeping, and giving priority to those connected with bee diseases. To date eight awards have been made to workers in six universities/research centres.

Donations in his memory will be most welcome by the Trust and should be sent to The Secretary, 104 Lower Luton Road, Wheathampstead, Herts. AL4 8HH.

CROSSWORD SOLUTION

E	A	S	Y		C	O	M	M	A	N	D	O
I		U		B		N		E		O		N
D	A	N	C	E		R	E	A	R	G	U	N
E				S		O		N		R		U
R	I	A		S	M	O	K	E	H	O	O	D
	B				F		S		U		E
B	Y	D	U	S	K		S	T	I	N	G	S
E		O		W		N			D	
J	U	M	P	A	H	E	A	D		S	I	D
A		I		L		C		U			O
S	I	N	G	L	E	T		N	O	B	B	Y
U		A		O		A		K		E		E
S	I	L	K	W	O	R	M		B	E	E	N

THE Boo Bees

FROM THE SILVEY-JEX PARTNERSHIP

Published in Great Britain by Silvey-Jex Publications
14 Chaldon Road, London   SW6 7NJ

DIRECTORY OF BEEKEEPING ORGANISATIONS & STATISTICS

All efforts have been made to ensure the accuracy of the information in these pages. Corrections and amendments should be sent to the Editor, Manor Farm, Upton, Newark, Nottinghamshire NG23 5ST

ASSOCIATIONS AND SERVICES

	PAGE
Apicultural Education Association (AEA)	96
Bee Diseases Association (BDI)	97
Bee Farmers' Association (BFA)	98
Beekeeping Courses and Services (BCS)	100
Beekeeping Editors' Exchange Scheme (BEES I & II)	102
Bees for Development (BFD)	103
British Beekeepers' Association (BBKA)	104
British Isles Bee Breeders' Association (BIBBA)	112
C.B. Dennis British Beekeepers Research Trust (CBDBBRT)	113
Central Association of Beekeepers (CABK)	115
County Beekeeping Magazines and Newsletters (BEEMAGS)	116
Council of National Beekeeping Associations of the United Kingdom (CONBA)	117
Devon Apicultural Research Group (DARG)	118
Federation of Irish Beekeepers' Associations (FIBKA)	119
International Bee Research Association (IBRA)	122
Institute of Apiculture (IOA)	124
National Beekeeping Periodicals (NATMAGS)	125
National Diploma in Beekeeping (NDB)	126
National Honey Show (NHS)	128
Queen Mary's College Westfield (QMCW)	129
Rothamsted Experimental Station (RES)	130
Scottish Beekeepers' Association (SBA)	132
Ulster Beekeepers' Association (UBKA)	137
Welsh Beekeepers' Association (WBKA)	138

BEEKEEPING STATISTICS FOR 1993 AND ADVISORY SERVICES

National Bee Unit (CSL)	143
Department of Agriculture Northern Ireland (DANI)	146
Scottish Agricultural Science Agency (SASA)	147
Scottish Area Offices (SOAFD)	150
Republic of Ireland (TEAGASC)	151
England and Wales Statistics (MAFF)	153

AEA

APICULTURAL EDUCATION ASSOCIATION

Secretary: *Clive de Bruyn* NDB,
Rural Education & Training Centre
Writtle College
Chelmsford, Essex
CM1 3RR
☎ (01245) 420705

Treasurer: *Margaret Thomas* NDB

The function of AEA is to act as a forum for professionals with an interest in Beekeeping Education to meet and exchange opinions and ideas.

Membership is open to those working in professional beekeeping education or allied areas, full or part time.

BEE DISEASES INSURANCE LTD.

Secretary: *Michael Wakeman*
                    Lower Goytre Farm
                    Knighton, Powys LD7 1UY
                    ☎ (01547) 510224

Manager, Scheme A:
                    Bernard White
                    Honeycroft, 36 Shaw Lane Gardens
                    Guiseley, West Yorkshire LS20 9JH
                    ☎ (01943) 879761

Manager, Scheme B:
                    Mrs. M. Homer
                    Eadie Cottage
                    Cutnall Green
                    Droitwich, Worcestershire WR9 0PQ
                    ☎ (01299) 851637

Bee Diseases Insurance provides insurance cover for individual beekeepers, association apiaries and commercial beekeepers alike, against the possibility of their bees and equipment being destroyed as a result of a Destruction Order being levied by a visiting officer from MAFF.

The most common disease resulting in destruction is American foul brood but insurance also covers European foul brood at the moment. It should be borne in mind that colonies infected with Varroasis will be weakened and are therefore more susceptible to infection by AFB.

Scheme A provides cover for the beekeeper with a total of 39 hives or less. Cover is usually obtained through a member's association although a group of eight or more people can also make an application for cover. Initial enquiries should be made through local associations which will refer the applicant to the Scheme A Manager.

Scheme B provides cover for beekeepers with 40 or more colonies in total. Insurance under this Scheme is on a personal basis and further details can be obtained from the Scheme B Manager.

In addition to providing insurance cover B.D.I. also supplies medicaments for the treatment of diseases. Supplies and further information are obtainalbe from the Manager, Scheme A.

REMEMBER: DISEASE CAN STRIKE ANY COLONY AT ANY TIME AND IT IS SPREAD THROUGHOUT THE COUNTRY. PROTECT YOUR APIARY THROUGH B.D.I.

BEE FARMERS' ASSOCIATION

OF THE UNITED KINGDOM

General Secretary:
B.A. Stenhouse
Borders Honey Farm
Newcastleton
Roxburghshire
TD9 0SG
☎ (01387) 376737

Chairman: J.D. Cossburn, Roselea, Hamdown Crescent, East Wellow, Romsey, Hampshire SO51 6BJ

Vice chairman: A. Chambers, 75 Branksome Hill Road, College Town, Camberley, Surrey GU15 4QF

Treasurer: M.J. Chandler, Dyers Hall Farm, Sundon Rd, Harlington, Dunstable LU5 6LL ☎ (01525) 874854

Pollination Sec.: M.R. Williams, 41 Clive Road, Sittingbourne, Kent ME10 1PJ ☎ (01795) 478800

BFA is affiliated to the National Farmers Union. Membership is restricted to commercial beekeepers with over 40 stocks. Business is conducted at two general meetings of members, the AGM held in London during the National Honey Show and the residential Spring Meeting held at various venues in March.

FUNCTIONS

- To monitor and to keep members informed about developments in commercial beekeeping, bee science and UK and EEC legislation.
- Liaison with Farmers, Growers, Contractors, Consumers and other organisations.
- Liaison with UK Government Depts dealing with beekeeping and allied matters.
- Liaison and co-operation with CONBA and UK Beekeeping organisations.
- Contact with European beekeeping organisations and representation on the EEC Honey Working Party (COPA/COGECA) in Brussels.
- Political lobbying through MPs and Euro MPs.
- To promote the production of high quality UK honey and to pursue cases of misleading labelling and misrepresentation of honey.
- To provide pollination services through the National Pollination Service.

FACILITIES FOR MEMBERS:

- Frequent Bulletins with news and updates, notes on meetings with MAFF and the EEC, reports on current beekeeping problems (e.g. varroa) and commercial developments.

- Free advertisement of members' sales and wants (including hive products, bee stocks and spare equipment).
- Crop and winter loss reports.
- Circulation among members of UK and foreign magazines.
- Free insurance for products liability and third party.
- Comprehensive special beefarmers insurance with the NFU.
- Pollination contracts.
- Advice from experienced members on all aspects of honey farming and commercial beekeeping; sources of equipment and sundries.
- Product directory listing specialist suppliers.
- Discounts from suppliers.
- Bulk purchase schemes.

BEEKEEPING COURSES AND SERVICES

FULL TIME COUNTY BEEKEEPING LECTURERS

These officers offer a comprehensive range of day and evening courses in both theoretical and practical beekeeping for those who wish to take up beekeeping and for experienced beekeepers. They also offer an adult bee disease diagnostic service.

Essex	*Clive de Bruyn* NDB, Rural Education & Training Centre, Writtle College, Chelmsford, Essex CM1 3RR ☎ (01245) 420705
Gloucestershire	*Tony Boonham*, Hartpury College, Hartputry House, Gloucester GL19 3BE
Kent	*Brian Palmer*, Hadlow College, Hadlow, Tonbridge TN11 0AL ☎ (01732) 850551
Surrey	*D.H. Daniels*, Merrist Wood Ag. College, Worplesdon, Guildford GU3 3PE ☎/ (01483) 232424

PART TIME LECTURERS & FURTHER EDUCATION COURSES IN BEEKEEPING

The following may offer a range of theoretical and practical courses in beekeeping, in some cases an advisory service or a diagnostic service for adult bee diseases only may be offered. The range of services and activities is wide and this list is not exhaustive but the following may be contacted for details of facilities in an enquirer's area.

Bedfordshire	*A.R.W. Griffin* NDB, 64 Leafields, Houghton Regis, Dunstable LU5 5LX
Berkshire	*George Butler*, Berks College of Agriculture, Burchetts Green, Maidenhead SL6 6QR
Devon	*Dr. Mick Street*, Bicton College of Agriculture, Budleigh Salterton EX9 7BY
Gloucestershire	*J.S. Cox*, Gloucestershire College of Agriculture, Hartpury House, Gloucester GL19 3BE
Hereford & Wor.	*Jim Crundwell* NDB, Pershore College of Horticulture, Worcester WR10 3JP
Hertfordshire	*Peter Dalby*, Oaklands College, Hatfield Road, St. Albans AL4 0JA
Humberside	*R. Chambers*, College of Agriculture, Bishop Burton, Beverley HU17 8QG
Norfolk	*Paul Metcalf* NDB, Easton College, Easton, Norwich NR9 8DX
Somerset	*David Charles*, Cannington College of Agriculture & Horticulture, Bridgwater TA5 2LS
Wiltshire	*Oliver Menhinick*, Lackham College, Lacock, Chippenham SN15 2NY

Suggestions for additions to this list are welcome

BEEKEEPING AT WRITTLE COLLEGE

Essex is one of the few counties which has a full time County Beekeeping Instructor whose job covers all aspects of beekeeping education. Advice to the general public on bees, beekeeping, hive produce, pollination and insecticide sprays is one aspect of the job, nightschool classes and dayschools for new and experienced beekeepers is another. Instruction for the full time college students and local schools and associations also comes within this brief. There are also opportunities for residential week long courses and long weekend breaks for new beekeepers and experienced apiarists.

A number of out apiaries are maintained for more advanced beekeeping courses and specialist activities such as queen rearing.

Courses:
A number of courses are run at various centres in the county. In the winter months the theoretical aspects of beekeeping are covered. Some of these classes are specifically geared to the BBKA examinations. In the summer, when the honeybees are active, practical beekeeping classes are run so that students can learn the techniques of handling bees.

Dayschools:
In the last 18 months dayschools have been run on the following topics:

- Queen rearing
- Pollen Trapping
- Candle making
- Honey judging
- Pollen supplement making and feeding
- Brood disease recognition
- Propolic collection and marketing
- Anatomy of the honeybee
- Computers in beekeeping

- Pollen analysis
- Skep making
- Royal Jelly
- Swarm control
- Honey marketing
- Foundation making
- Varroa update
- Microscopy

For Further information on any classes, courses or dayschools contact:

Clive de Bruyn
(County Beekeeping Instructor)
Writtle College
Chelmsford
Essex CM1 3RR
☎ (01245) 420561

BEEKEEPING EDITORS' EXCHANGE SCHEME

Contact: *Mary Fisher*
38 St Martin's View
Leeds   LS7 3LB
☎ (0113) 262 1201

BEES I & II are self-help associations of local, county and country association beekeeping editors which operate principally by exchanging journals monthly through a central address. The schemes are supported by Northern Bee Books.

BEES I was founded in 1984, BEES II in 1994, both after successful beekeeping ediors' courses when it became evident that editors frequently do not have the support they need from their associations. Now fully established as part of the British, Irish Republican and British Columbian beekeeping scene the schemes can make participants and their members the best informed and most up to date beekeepers anywhere. Enthusiasm for BEES is confirmed by a nil drop-out rate.

The aims are:

• to exchange ideas for content and production methods
• to aid others by experience
• to communicate matters editorial
• to share information on national beekeeping issues
• to help and reassure those new to the task
• to give a wider readership to the best writing in
  beekeeping journalism

If you are an editor or potential editor and would like to know more about how we operate write to Mary Fisher.

BEES *FOR* DEVELOPMENT

Contact: *Dr. Nicola Bradbear*
Bees *for* Development
Troy
Monmouth NP5 4AB
☎ (01600) 713648 **Fax:** (01600) 716167

Bees *for* Development encourages people to practise sustainable beekeeping. It is a not-for-profit organisation providing a friendly information service on all aspects of bees and beekeeping in the poorest countries of the world.

Bees *for* Development maintains an active network between interested people.

Communication and discussion is carried in the journal *"Beekeeping and Development"*. This fully illustrated quarterly is packed with very different beekeeping information. It provides ideas for low-technology beekeeping, vital contacts, details of activities, new publications and current news of the beekeeping world.

We need beekeepers everywhere to support our efforts. You can help us by:

- Subscribing to *"Beekeeping and Development"*. Subscriptions to this journal help us to continue our information service for beekeepers in developing countries. The journal costs £12 per year in the UK. Special rates are also available for bulk subscriptions by beekeeping associations;

- Sponsoring additional subscriptions - we welcome sponsored subscriptions and so do the recipients!

- Sharing your expertise;

- Buying your bee books from us. Bees *for* Development offers a mail order book service. All income we generate from book sales is used towards providing information to beekeepers in developing countries. Ask for our book list, *Books to Buy"*.

Write to us, or call us, at the address above.

BRITISH BEEKEEPERS' ASSOCIATION

General Secretary:
> *Adrian C. Waring*
> BBKA Headquarters
> NAC, Stoneleigh Park
> Warwickshire   CV8 2LZ
> ☎ (01203) 696679 **Fax:** (01203) 690682

BBKA Headquarters:
> *National Beekeeping Centre,* National Agricultural Centre,
> Stoneleigh, Kenilworth   CV8 2LZ   ☎ (01203) 696679 **Fax:** (01203) 690682
> (Office hours 9.00am-5.00pm Monday, Tuesday & Friday.
> Telephone answering service outside office hours)
> *Mrs. Sally Edwards,* Assistant Secretary

EXECUTIVE COMMITTEE 1995/1996

President: *J.D. Frimston* LLB NDB, Brockworth, Tower Road, Heswall, Wirrall L60 6RT

Chairman: *A.R. Johnson,* Grasmead, Dean, Bishops Waltham, Southampton   SO32 1FY

Vice Chairman: *Gerald Moxon,* 9 Savery Street, Hull   HU9 3BG

Members: *M.J. Badger,* Kara, Thorn Lane, Roundhay, Leeds   LS8 1NN
P. Beckley, Glebelands, Star Lane, Rockland St. Mary, Norfolk   NR14 7BX
A.J. Boonham, Hartpury College, Hartpury, Nr. Gloucester   GL19 3BE
E.R. Chapman, Hill Rising, Corsley, Warminster, Wiltshire   BA12 7PG
Mary Dartnall, 2 Harlyn Road, Southampton   SO1 4NF
B.P. Dennis, 50 Station Road, Cogenhoe, Northamptonshire   NN7 1LU
Mrs. A. McKenzie, 99 Harbourne Road, Edgbaston, Birmingham   B15 3HG
W.S. Munday, 34 King Edward Avenue, Dartford, Kent   DA1 2HY
Barry Potter, 29 Knapton Lane, Acomb, York   YO2 5PX
I. Preece, Appledore, Box View, Colerne, Nr. Chippenham, Wilts.   SN14 8DH

Co-opted Members:
J.D. Yates, (Chairman of the Examinations Board),
Tides Reach, Yealm View Road, Newton Ferrers, Plymouth   PL8 1AN

Treasurer: *Leslie Hewitson* FCII, 48 Derwent Drive, Purley, Croydon   CR8 1EQ

Bee Craft Rep: *D. Ribbans,* 48 Castle Road, Whitstable, Kent   CT5 2DY

COMMITTEES OF THE EXECUTIVE AND SECRETARIES

ANNUAL DELEGATE MEETING AND FINANCE
General Secretary

EDUCATION & HUSBANDRY
The Husbandry Committee also covers bee diseases and European & UK Government regulations. Details from the Secretary, *R. Chapman*

EXAMINATIONS BOARD
> *John Hendrie*, 26 Coldharbour Lane, Hildenborough, Tonbridge, Kent TN11 9JT ☎ (01732) 833894 (home) (01903) 820692 (Office)

TECHNICAL
The Technical Committee covers: Environmental Issues, Research and Scientific, Spray and Pesticides matters of concern. Details from the Secretary, *A.C. Waring*

SUBSCRIPTIONS AND MEMBERSHIP FEES
Member Associations fix their annual subscription rates according to their financial requirements and the services they provide for their members. The annual capitation fee payable by County and Area Member Associations to BBKA is £5.00 for 1994/5. The membership fee payable by Specialist Member Associations is £25 per annum and the Individual Members' subscription is £10 per annum, overseas £12 per annum.

EXAMINATIONS
The Examinations Board of BBKA performs a national function, providing a structured range of examinations fulfilling the needs of all beekeepers. All matters concerning examinations, except for the correspondence course, should be addressed to the Examinations Secretary.

CORRESPONDENCE COURSES
Courses to prepare candidates for the Basic Examination, Intermediate Examination and the Senior Examination, Part 2, are available.
Details from the Courses Secretary, *G.M. Collins*, 72 Tatenhill Gardens, Doncaster DN4 6TL.

LEGAL ADVICE
The Legal Adviser to the BBKA may be able to help Local Associations with legal problems to a limited extent. Contact through the General Secretary.

SPRAY PROBLEMS
Help and advice for members who have suffered losses from poisonous sprays may be obtained from the Technical Committee Secretary. A leaflet, **'Protecting Honey Bees from Pesticides'** is available from Headquarters.

EVENTS
The various gatherings of beekeepers continue to be a feature of BBKA's many functions and provide a vital service for the dissemination of knowledge and the drawing together of members.

The three day Convention held in conjunction with the National Honey Show and the Spring Convention at Stoneleigh, combined with the comprehensive features of the Bee Fair, are now firmly established events. From time to time day or weekend courses are held on special subjects. The BBKA Dinner is held at the start of the Central Association Conference at Leamington Spa. For the past three years there has been a stand featuring bees and beekeeping at the Chelsea Flower Show.

INSURANCE

Every Individual Member of BBKA and every fully paid-up member of a Beekeeping Association affiliated to BBKA is indemnified under London and Edinburgh Insurance plc Public Liability policies for sums which they may become legally liable to pay in respect of claims arising out of beekeeping activities made against them for injury or for damage to property. The cover extends to maximum of £2,000,000 arising from any one accident or series of accidents. The beekeeper must bear the first £100 of each and every claim.

Two other types of insurance are available to beekeepers:

1. The *"All Risks"* which covers loss of or damage to hives and equipment
2. The BBKA Combined Insurance Scheme for Member Associations, which covers various activities of associations.

Further details can be obtained from the BBKA Treasurer.

EDUCATION COMMITTEE

This committee was established to develop apicultural education and the BBKA correspondence courses.

PUBLICATIONS

- *Bee Craft* is the monthly official journal of BBKA. Subscription is £8.96 p.a. including postage (new subscribers special rate £8); overseas subscription £12.96. Orders to *Mrs. S. White*, 15 West Way, Copthorne Bank, Crawley   RH10 3QS   ☎ (01342) 712119.
    There are special rates for associations, details from Mrs. White.
- *BBKA Year Book* contains detailed information about BBKA including the Annual Report of the Executive and other committees and the Annual Accounts. Copies can be obtained from Headquarters for £1.00p + 9" x 6" s.a.e.
- *BBKA News* is issued six times a year, free to all members of BBKA. The Editor/Advertising Manager is *Mrs. S. Blake*, Stratton Court, Over Stratton, South Petherton, Somerset   TA13 5LQ Distribution is organised by *Mary Dartnall*, 2 Harlyn Road, Southampton   SO1 4NF
- *Association Officers' Handbook* is about running an association and contains useful information for all area and local BKA officers. £2.00 from BBKA HQ.
- *Directory of Lecturers & Demonstrators* - compiled by the Education Committee to help those who organise meetings. Send 60p and a 9" x 6 $^1/_2$" s.a.e. to H.Q.

ITEMS FOR SALE

The Association's range can be bought at several of the major national events.

Mail order enquiries should be addressed to *Mrs. G. Chirnside*, Bryn y Pant Cottage, Upper Llanover, Abergavenny   NP7 9ES.

The main items are:

- **Leaflets** - a number of reasonably priced leaflets on beekeeping matters.
- **Show material** - a range of Prize Cards, Class Cards, Labels and specimen Show Rules, Honey Grading Filters and Medals.
- **Hive Plans** - detailed plans for the construction of Modified National, Smith, WBC hives and the Sundown Floor.
- **Other items** - Honeybee Anatomy Transparencies, Badges, Brooches, Ties. An up-to-date 30 min video on varroa.

AUDIO-VISUAL AIDS LIBRARY

A wide selection of video cassettes, projection slides and sound cassettes on beekeeping and conservation generally is available, including two videos on varroa.

Requests for video and sound cassettes should be made to BBKA HQ, requests for projection and microscope slides should be made to BBKA Slides Officer J.W. *Rainey*, Delft Cottage, Wood Lane, Aspley Guise, Milton Keynes MK17 8EJ ☎ (01908) 582888.

Catalogues of available material are available for 75p on receipt of a 9" x 6" s.a.e. New material is reviewed by the librarians who welcome enquiries, suggestions and donations of items to improve the service.

NATIONAL BEEKEEPING CENTRE

BBKA Headquarters Office, Bee Garden and Pavilion are on a site in the Royal Agricultural Society of England's Showground at Stoneleigh, Warwickshire. The Bee Garden of about one acre contains plants and shrubs of interest to bees, two observation hives, specimen hives of various designs and two bee boles. The six hive apiary is managed throughout the year by the Stoneleigh apiarists and is available to Associations for demonstrations.

The Dixon Memorial Pavilion is used for educational and training seminars, honey show displays during the Royal Show (July) and The Town and Country Exhibition (August). It may be booked for Association meetings. The Bee Centre and Gardens are open to the general public during all major events organised by RASE and during office hours. The object of RASE is to present agriculture and allied industries to the world. BBKA is part of this presentation; close liaison exists between BBKA and RASE and is an indication of continued co-operation.

MEMBER ASSOCIATIONS AND THEIR SECRETARIES

Avon Dr. I. *Davis*, "Landi Kotal", Brinsea Road, Congresbury BS19 5JJ
☎ (01934) 832825

Berkshire G.W. *Knights*, 28 Argyle Street, Reading RG1 7YP ☎ (01734) 596979

Bournemouth Mrs. E. *Ware*, 23 Barrowgate Road, Throop, Bournemouth, Dorset

Bucks. Mrs. F. *Pigram*, 4 Hobart Close, High Wycombe HP13 6UF ☎ (01494) 523783

Cambridgeshire N.T. *Heywood*, Owls Wood Cottage, New Sells Village, Royston, Herts.
SG8 8DE ☎ (01763) 848801

Cheshire M.F. *Haynes*, 98 Gatley Road, Gatley, Stockport SK8 4AB ☎ (0161) 491 2382

Chesterfield Mrs. S. *Wright*, 21 Mitchell Street, Clowne, Derbyshire S43 4SH
☎ (01246) 810771

Cornwall R. *Walker*, 8 Lower Brae, Brae, Camborne, Cornwall TR14 8NZ
☎ (01208) 872300

Cornwall West Mrs. L. *Bryning*, Roseladden, Sithney, Nr. Helston, Cornwall

Cumbria J. *Skinner*, 25 Beech Lane, Cockermouth, Cumbria CA13 9HQ
☎ (01900) 823270

Derbyshire Mrs. M. *Cowley*, 14 Montpelier, Quarndon, Derby DE6 4JW ☎ (01332) 556227

Devon C. *Utting*, Codden, Golf Links Road, Westward Ho!, North Devon
EX391HH ☎ (01237) 474500

Dorset R. *Norman*, 19 Broughton Crescent, Wyke Regis, Weymouth, Dorset DT4 9AS
☎ (01305) 786585

Dover & Dist. Miss E. *Gordon*, 10 Radnor Park Road, Folkstone, Kent ☎ (01303) 251971

Durham	*V.I. Collinson*, Hillside Bungalow, Rushyford, Ferryhill, Co. Durham ☎ (01388) 721345
Essex	*Mrs. J. Smye*, 8 Gate Street Mews, Maldon, Essex CM9 7EF ☎ (01621) 850605
Glos.	*D. Streatfield*, Corderies, Down Road, Alveston, Thornbury, Bristol BS12 2JE ☎ (01454) 413258
Gwent	*P. Hayward*, Llananant Farm, Penallt, Monmouth NP5 4AP ☎ (01600) 712864
Hampshire	*Capt. A.R. Johnson*, Grasmead, Dean, Bishops Waltham, Southampton SO3 1FY ☎ (01489) 892390
H'gate & Ripon	*J. Stephenson*, 4 Brookfield, Hampsthwaite, Harrogate, North Yorkshire HG3 2EF ☎ (01423) 770736
Herefordshire	*Mrs. H.L. Beaver*, Rose Cottage, Upper Hill, Leominister HR6 0JZ ☎ (01568) 720343
Hertfordshire	*Mrs. O.M. Gabriel*, 25 Winton Drive, Cheshunt, Herts. EN8 9JP ☎ (01992) 626903
Hunts.	*Mrs. R. Backhouse*, 2 Orchard Close, Elsworth, Cambridge CB4 5HG ☎ (01954) 267342
Isle of Man	*Miss D. Cringle*, Barrule Cottage, St. Marks, Ballasalla ☎ (01624) 825564
Isle of Wight	*Mrs. J. Starling*, Afton, Cranmore Avenue, Yarmouth PO41 OXS ☎ (01983) 760634
Kendal & S. W'ld	*J.M. Bottomley*, 21 Hillcrest Drive, Slackhead, Beetham, Milnthorpe LA7 7BB ☎ (01539) 562084
Kent	*Dr. J.W. Cowan*, 121 The Grove, West Wickham, Kent BR4 9LA ☎ (0181) 777 1841
Kings Lynn	*Mrs. M. Dunkley*, Castle Farmhouse, Wormegay, Kings Lynn PE33 0SE ☎ (01533) 840315
Lancs. & N.W.	*Mrs. L. Hogarth*, Wyreside Filling Station, Garstang Road, St. Michael's-on-Wyre PR3 0TD ☎ (01995) 679631
Leics. & Rut'd	*F.B. Cramp*, 2 Woodland Drive, Groby, Leicester LE6 0BQ ☎ (0116) 287 6879
Lincolnshire	*Mrs. A. Holderness*, 15 Holdingham, Sleaford, Lincs. NG34 8NR ☎ (01529) 302774
London	*N. Morris*, 34 Albion Drive, London E8 4ET ☎ (0171) 249 6800
Ludlow & Dist.	*Mrs. M. Weston*, Blackwell Cottage, Clun, Nr. Craven Arms, Shropshire ☎ (0158) 84 723
Manch. & Dist.	*Mrs. M. Bohme*, 54 Dunster Drive, Flixton, Manchester M31 3WR ☎ (0161) 747 7292
Middlesex	*Mrs. J.V. Telfer*, Midwood House, Elm Park Road, Pinner HA5 3LH ☎ (0181) 868 3494
Mole Apiary Club	*D. Cutler*, Romney House, 70 Hurst Road, East Molesey, Surrey KT6 1AL ☎ (0181) 979 1423
N'castle & Dist.	*C. Burn*, 19 Lanercost Drive, Fenham, Newcastle-upon-Tyne NE5 2DH ☎ (0191) 274 4932
Norfolk	*H. Copperthwaite*, 1 The Maltings, Millgate, Aylsham, Norfolk NR11 6HX ☎ (01263) 734682
Northants.	*D.A. Ward*, 109 Bouverie Rd., Hardingstone, Northampton NN4 0EG ☎ (01604) 765454
Northumberl'd	*Dr. R. Lowther*, Old School House, Wall Village, Hexham NE46 4EF ☎ (01434) 681660
Notts.	*A. Barber*, 11 Old Hall Gardens, Coddington, Newark, Notts. ☎ (01636) 71844
Oxfordshire	*R.H. Redman*, 192 Poplar Grove, Kennington, Oxford OX1 5QT ☎ (01865) 739814

Peterborough	*P.G. & J.E. Newton*, 65 Queen Street, Yaxley, Peterboro' PE7 3JE ☎ (01733) 243349
Shropshire	*Mrs. P. Mills, 25 Collett Way, Telford TF2 9SL* ☎ (01952) 293621
Shropshire N.	*C. Wingett*, Greenwich House, 3 The High Street, Overton-on-Dee, Clwyd LL13 0DT ☎ (01978) 710294
Somerset	*Mrs. B. Quartly*, Jeanes Bungalow, Bradford-on-Tone, Taunton TA4 1AX ☎ (01823) 461748
Staffs. N.	*Mrs. K.E. Silver*, "Oak Ridge", Gratton, Nr. Endon, Stoke-on-Trent ST9 9AQ ☎ (01782) 502380
Staffs. South	*D.J. Clift*, Rose Cottage, Long Lane, Newtown, Great Wyrley, Staffs. WS6 6AU ☎ (01922) 401105
S'ford on Avon	*D.N. Keyte*, Sunnybank, Wootton Wawen B95 6BH ☎ (01564) 792872
Suffolk	*J.D. Quinlan*, The Old Rectory, Dallinghoo, Woodbridge, Suffolk ☎ (01473) 737700
Surrey	*Mrs. E. Mence*, 27 Acacia Grove, New Malden KT3 3BJ ☎ (0181) 942 7505
Sussex	*S.D. Kelly*, 31 Ferrers Road, Lewes, East Sussex BN7 1PY ☎ (01273) 477164
Sussex West	*P. Stoehr*, Piper Lodge, 22 Cottingham Avenue, Horsham, West Sussex RH12 4HU ☎ (01403) 261259
Thanet	*T.K. Williams*, 31 Alpha Road, Birchinton CT7 9EG ☎ (01843) 42050
Twickenham & Thames Valley	
	P. Crowther, 13 Cranbrook Drive, Whitton TW2 6HN ☎ (0181) 894 9275
Warwickshire	*Mrs. C.F. Davis*, The Pines, Hidgetts Lane, Berkswell CV7 7DG ☎ (01676) 533252
Wiltshire	*A.K. Ludlow,* 13 The Paddock, Urchfont, Devizes, Wiltshire SN10 4SH ☎ (01380) 848213
Worcestershire	*Mrs. U. Brandwood,* 10 Monnow Close, Droitwich Spa WR9 8TF ☎ (01905) 772070
Wye Valley	*C.W. Knowles*, Nash Hill Cottage, Kilpeck, Herefordshire HR2 9DW ☎ (01981) 21 238
York & Dist.	*T. Robinson*, 71 Broadway, York YO1 4PJ ☎ (01904) 626170
Yorkshire	*G.M. Moxon*, 9 Savery Street, Southcoates Lane, Hull HU9 3BG ☎ (01482) 782052

UNAFFILIATED BEEKEEPING ASSOCIATIONS

Bedfordshire	*R. Sherwood*, 157 Bedford Road, Wootton MK43 9BA
Carshalton	*H.C.J. Steggals*, 111 Carshalton Park Road, Carshalton Beeches, Carshalton SM5 3SJ
Enfield	*D.L. Curtis*, 47 Hadrians Ride, Lincoln Road, Enfield EN1 1DF
Medway Co. Kent	
	P. Griffiths, Park House, The Street, Newnham, Sittingbourne ME9 0LU
Jersey	*E. Gautier*, Cromwell Lodge, Belvedere Hill, St. Saviour, Jersey, Ch. Islands

ASSOCIATION EXAMINATION SECRETARIES

Avon	*G. Simms*, Westholm, Lake Road, Portishead, Bristol BS20 9JA
Berkshire	*G.R. Hawthorne*, 69 West Chiltern, Woodcote, Reading RG8 0SG
Bucks.	*D.J. Pigram,* 4 Hobart Close, High Wycomb, Bucks. HP13 6UF ☎ (01494) 523783
Derbyshire	*B. Fletcher,* 217 Henhurst Hill, Burton-on-Trent, Staffordshire DE13 9SX

Devon	*J.D. Yates*, Tides Reach, Yealm View Road, Newton Ferrers, Plymouth, Devon PL8 1AN
Dorset	*R.O.M. Page*, 452 Chickerell Road, Weymouth DT3 4DH
Dorset South	*Mrs. M. Davies*, 80 Leybourne Avenue, Ensbury Park, Bournemouth, Dorset BH10 6HE
Essex	*W.D. Fildes*, 18 Andersons, Stanford-le-Hope SS17 7JF
	D.J. Webber, Breewood Cottage, School Lane, Great Horkesley, Colchester, Essex CO6 4BW
Glos.	*K.B. Durk*, 242 London Road, Charlton Kings, Cheltenham GL52 6HS
Hampshire	*J.D. Cossburn*, Roselea, Hamdown Crescent, West Wellow, Romsey
H'gate & Ripon	*M.J.M. Annett*, 3 Rossett Drive, Harrogate HG2 9NS
Herefordshire	*P. Quilliam*, 71 Seaton Avenue, Hereford HR1 1NP
Hertfordshire	*A.P. Dalby*, 37 Cecil Road, Cheshunt, Herts. EN8 8JH
Kent	*Mrs. J.S. Lea*, Maypole Farm, Maypole Lane, Goudhurst, Kent TN17 2QP ☎ (01580) 211614
Lancashire	*C. Rawnsley*, 31 Brunswick Road, Morecambe, Lancashire
Lincolnshire	*B.C. Whaler*, Smithills, Scothern Lane, Sudbrooke LN2 2QJ
Oxfordshire	*Mrs M. Rees*, 4 Quarry Hollow, Headington OX3 8JR
Notts.	*D. Guthrie*, 9 Winchilsea Avenue, Newark, Nottinghamshire NG24 4AD
Shropshire N.	*Mrs. A. Thorpe*, Wynn Cottage, Lower Perthy, Ellesmere, Shropshire SY12 0HX ☎ (01691) 622852
Somerset	*David Charles*, Bickerton, Church Lane, West Pennard, Glastonbury BA6 8NT
Staffs. N.	*Dr. N.C. Mawby*, Glenwood, Wood Lane, Longsdon, Stoke on Trent ST9 9QB ☎ (01538) 399322
Staffs. S.	*D.J. Clift*, Rose Cottage, 14 Long Lane, Newtown, Great Wryley WS6 6AU ☎ (01922) 401105
Suffolk	*Mrs. Sally Green*, 22 Old Street, Haughley, Stowmarket, Suffolk IP14 3HX ☎ (01449) 770391
Sussex East	*Mrs. D. McKecknie*, 137 High Street, Lindfield, Haywards Heath, West Sussex RH16 2HR
Sussex West	*R. Hamilton*, 137 Pondtail Road, Horsham RH12 5HT
Warwickshire	*P.D. Lishman*, Aston Farm House, Newtown Lane, Shustoke, Coleshill B46 2SD
Worcs.	*D.P. Friel*, 17 Tennal Road, Harborne, Birmingham B32 2JD
Yorkshire	*G.M. Collins*, 72 Tatenhill Gardens, Doncaster, S. Yorkshire DN4 6TL

Where Associations have no Examinations Secretary the Association Secretary deals with examinations. To help future candidates it is suggested that Associations without an Examination Secretary appoint one. Associations are responsible for arranging a suitable room for the written Examinations and recommeding an invigilator.

ACTIVE HOLDERS OF THE BBKA SHOW JUDGES CERTIFICATES

Badger M.J.	14 Thorn Lane, Leeds LS8 1NN
Blackburn, Mrs. H.M.	
	15 Highdown Hill Road, Emmer Green, Reading RG4 8QR
Brown Mrs. V.	20 Swains Lane, Flackwell Heath HP10 9PU
Capener Rev. H.F.	1 Baldric Road, Folkestone CT20 2NR
Clark, Mrs. R.E.	10 The Park, Bookham, Surrey KT23 3JL
Collins G.M.	72 Tavenhill Gardens, Doncaster DN4 6TL

Cooper Miss R.M.	10 Gaskells End, Tokers Green, Reading RG4 9EW
Creighton R.	Long Reach, Stockbury Valley, Sittingbourne ME9 7QP
Creighton Mrs. K.	Long Reach, Stockbury Valley, Sittingbourne ME9 7QP
Daniels D.H.	Twin Oaks, Nightingale Road, Ash, Aldershot, Hants.
Davies Mrs. M.	80 Leybourne Avenue, Ensbury Park, Bournemouth BH10 6HE
Diaper B.	69 Russell Bank Road, Four Oaks, Sutton Coldfield B74 4RQ
Dickson, Ms. F.	Didlington Manor, Didlington, Thetford, Norfolk IP26 5AT
Duggan R.M.	Redstone Wood Cottage, Philanthropic Lane, Redhill RH1 4DF
Fielding L.G.	Linley, Station Road, Lichfield WS13 6HZ
Hart H.R.	8 Manor Rise, Bearsted, Maidstone ME14 4DB
Hawthorne G.R.	69 West Chiltern, Woodcote, Reading RG8 0SG
Hender L.	Higher Cottage, Westowe, Lydeard St. Lawrence, Taunton TA4 3SH
Hughes E.	5 Rowdale Crescent, Sheffield S12 4SJ
Jones W.J.	Avalon, Llanfrothen, Gwynedd LL48 6LJ
MacGiollaCoda M.C.	
	Glengarra Wood, Burncourt, Cahir, Co. Tipperary, Republic of Ireland
Moore M.	4 Beechgrove, Athy, Co. Kildare, Republic of Ireland
McCormick E.	14 Akers Lane, Eccleston, St. Helens, Lancs. WA10 4QL
Nichols L.A.	Lowlands, Nats Lane, Brook, Ashford, Kent
Orton J.	Occupation Road, Sibson, Nuneaton CV13 6LD
Rawlings A.	The Elms, Frog Lane, Upper Boddington, Daventry NN11 6DJ
Read A.J.	Everest, Firs Road, Winterslow, Salisbury SP5 1ST
Rodgers J.S.	46a Turner Road, Mile End, Colchester CO4 5LA
Rolt Mrs. R.E.	Ambonne, Northfields Lane, Aldingbourne, Chichester PO20 6UH
Rounce J.N.	Mill View, Scarborough Road, Great Walsingham NR22 6AB
Scruby J.T.W.	Pilgrims Ridge, Markway, Godalming, Surrey
Sommerville G.	Valley View, Healey, Masham, Ripon HG4 4LH
Symes C.J.	189 Marlow Bottom Road, Marlow SL7 3PL
Taylor A.J.	The Old Pyke Cottage, Hethelpit Cross, Staunton GL19 3QJ
Tonsley C.C.	46 Queen Street, Geddington, Kettering NN14 1AZ
Tucker J.C.	18 Cedar Road, Woodsmoor, Stockport SK2 7DN
Vickery R.G.L.	Ponderosa, Verwood Road, Three Legged Cross, Wimborne BH21 6RN

HOLDERS OF THE BBKA ASSOCIATE JUDGES CERTIFICATE

Chirnside L.	Bryn-y-Pant Cottage, Upper Llanover, Abergavenny NP7 9ES
Donoghue J.	Bunnagappa, Walsh Islands, Geasehill, Co. Offaly, Eire
Duffin J.M.	Ringwood, Hampshire
Guest S.B.	Bridge House, Hindheath Road, Wheelock, Sandbach, Cheshire CW11 9LY
Harries C.D.	53 West View, Creech St. Michael, Taunton TA3 5DU
Lishman P.D.	Aston Farm House, Newtown Lane, Shustoke, Coleshill B46 2SD
Moxon G.	9 Savery Street, Hull HU9 3BG
Riches Dr. H.R.C.	2 South Approach, Moor Park, Northwood, Middlesex HA6 2ET
Salter T.A.	74 Hawthorne Road, Kings Norton, Birmingham B30 1EG
Smith P.L.	11 Hamilton Road, High Wycombe HP13 5BW
Thomas K.	Sir Benfro, 4 David Cross, Braunton EX33 2AT
Tipping J.M.	44 Gotherage Lane, Romiley, Stockport SK4 4AE
Weston B.	17 Downs Road, Penendenheath, Maidstone, Kent
Worthington K.G.	13 Greenfield Road, Bollington, Macclesfield, Cheshire SK10 5WE

BIBBA

BRITISH ISLES BEE BREEDERS' ASSOCIATION

Secretary: *Albert Knight*
11 Thomson Drive
Codnor Ripley
Derbyshire   DE5 9RU

Membership Secretary:
Brian Dennis, 50 Station Road, Cogenhoe, Northants   NN7 1LU
Publication Sales:
Alan Hinchley, Fold House Farm, Mickley Lane, Stretton, Alfreton,
Derbyshire   DE55 6FW
Breeding Groups Secretary:
Paul Arthur, 14 Whitbarrow Road, Lymm, Cheshire   WA13 9AF

BIBBA is an organisation devoted to encouraging beekeepers to breed native bees. The bee more suited to our environmental circumstances than other sub species. BIBBA's aims are publicised through books, workshops, lectures and conferences.

Breeding techniques advocated include:

• Assessment of colonies by observation, recording certain criteria on specially designed record cards.
• Determination and purity of sub species by measurement of morphometric characters
• Use of mini nucs for the mating of queens economically

BIBBA Publications include:

• *The Honeybees of the British Isles* by *Beowulf Cooper*
• *Breeding Techniques and Selection for Breeding of the Honeybee* by *Prof. F. Ruttner*
• *The Dark European Honey Bee* by *Prof. F. Ruttner, Eric Milner* and *John Dews*
• *Breeding Better Bees using Simple Modern Methods* by *John E. Dews* and *Eric Milner*

BIBBA encourages the formation of breeding groups, and the sharing of knowledge between groups.

For further information about BIBBA contact one of those named above.

THE C.B. DENNIS BRITISH BEEKEEPERS' RESEARCH TRUST

Registered Charity No. 328685

Hon. Secretary: *Ms. B.V. Ball*
104 Lower Luton Road
Wheathampsted
St. Albans
Herb. AL4 8HH

AIMS

This Charitable Trust was established in 1990 through the generosity of Mr. C.B. Dennis. It aims to use the interest from the capital investment of £65,000 to support British research projects that are likely to benefit beekeeping in the relatively short term, giving some priority to work on bee diseases. The total amount of income available for award annually is only about £5,000. This will allow the Trust to help existing projects or support short term investigations but not to initiate new areas of research.

The Trust is an independent body making awards to institutions or individuals on the basis of scientific merit of submitted proposals and perceived benefit to British beekeeping. Eight awards have been made to date and the following examples illustrate the diversity honey bee research supported.

Dr. Ingrid Williams of Rothamsted Experimental Station was assisted in the collection and analysis of data on pollination practice in the UK; information that is rarely published but essential for apiculture to be recognised as a vital component of sustainable agriculture in the nineties and beyond. The results of this work were presented in the lecture Dr. Williams gave at the BBKA Spring Convention in April 1992 and have now been published as part of wider studies.

Dr. William Kirk of the University of Keele, Staffordshire received support over three years to produce an updated and comprehensive colour guide to the pollen loads of honey bees in Britain. This practical aid to beekeepers has now been published and is an essential text for those who wish to identify the pollen sources of their bees.

Other awards have been made to Mark Allen of Rothamsted Experimental Station who aimed to develop a rapid, specific and sensitive serological test for detecting honey bee viruses in the parasitic mite *Varroa jacobsoni*, to the International Bee Research Association for portable computer equipment that could be used to increase the availability of their extensive databases to the beekeeping public at meetings and shows and to Fani Hatjina, a student at Cardiff University, to enable her to complete her studies on means of improving the efficiency of pollination by honey bees.

Application forms for research project funding and guidance notes for applicants are available on request from the Secretary.

DONATIONS

The Trust is seeking additional financial support to build up the investment capital so that sufficient income is generated to initiate the research that beekeepers would like to see undertaken. To achieve this end will take many donations of all sizes from individuals, branches, county associations and other organisations. All donations, however small, will be added to the capital and bee reasearch will benefit from the income in perpetuity. There are three ways you can help:

1. Make a small donation now

2. Persuade your branch or county association to make a larger donation

3. Consider leaving a bequest to the Trust in your will. As the Trust is a registered Charity this can reduce your inheritance tax liability. The Trust can accept shares or property as well as financial bequests.

Please think about this and help if you can.

All donations and correspondence should be sent to the Secretary.

THE CENTRAL ASSOCIATION OF BEEKEEPERS

Secretary:	*Margaret English* 6 Oxford Road Teddington TW11 0PZ
President:	*D.J. Little*, Little Greengarth, 60 Deben Avenue, Martlesham Heath, Ipswich, Suffolk IP5 7QP
Chairman:	*A.W. Ferguson*, IACR-Rothamsted, Harpenden, Herts. AL5 2JQ
Vice chairman:	*M.A. De Silva*, 31 Rosemary Avenue, Hounslow, Middlesex TW4 7JQ
Treasurer:	*T.B. Trood*, Blackdown Cottage, Curland, Taunton, Somerset TA3 5SA
Programme Secretary:	
	N.L. Carreck, IACR-Rothamsted, Harpenden, Herts. AL5 2JQ
Publications:	*R. Shead*, 11 Fairlight Road, Hadleigh, Benfleet SS7 2QJ

The Central Association of Beekeepers in its present form dates from the time of the reorganisation of the British Beekeepers' Association in 1945. The BBKA was originally made up of private members only. However as County Associations were formed they applied for affiliation and were later permitted to send delegates to meetings of the Central Association, as the private members were then known. This arrangement became unsatisfactory as the voting power of the Central Association greatly outnumbered that of the County Associations and so in 1945 a new Constitution was drawn up whereby the Council comprised Delegates from the Counties and Specialist Member Associations. The private members then formed themselves into a Specialist Member Association with the designation '**The Central Association of the British Beekeepers' Association**'; this was later shortened to its present style.

The Association was able to devote itself to its own particular aims, to promote interest in current thought and findings about beekeeping and aspects of entomology related to honey-bees and other social insects. Lectures given by scientists and other specialists are arranged, printed and circulated to members, as has been done since 1879.

Meetings are held in London at the Linnean Society, Burlington House, Piccadilly W1, with an annual Conference in Leamington Spa in October. The subscription is £6.00 per annum, £6.50 for joint membership (one copy only of publications).

COUNTY BEEKEEPING MAGAZINES AND NEWSLETTERS

Avon	*Brenda Davies*, Callow Cottage, Cuck Hill, Shipham, Winscombe BS25 1RD
Bedfordshire	*Mr. Hall*, 54 Marston Gardens, Luton, Bedfordshire
Berkshire	*George Hawthorne*, 69 West Chiltern, Woodcote, Reading RG8 0SG
Cambridgeshire	*Arthur Norman*, 17 Luard Road, Cambridge CB2 2PJ
Cheshire	*Jim Tucker*, 18 Cedar Road, Woodsmoor, Stockport SK2 7DN
Cornwall	*Gillian Searle*, 6 Harleigh Road, Bodmin, Cornwall
Cumbria	*John Skinner*, 25 Beech Lane, Cockermouth, Cumbria
Derbyshire	*Mrs. M. Cowley*, 14 Montpelier, Quorndon, Derbyshire DE22 5JW
Devon	*Peter Rosenfeld*, 42A Clifford Street, Chudleigh TQ13 0LE
Dorset	*Margaret Massey*, Burley, Shorts Green Lane, Motcombe, Shaftesbury
Durham	*Chris Appleby*, 24 Oak Rd., Easington, Peterlee SR8 3HR
Essex	*Jo Prentice*, Yeomans, Rolphy Green, Pleshey, Chelmsford CM3 5JQ
Gloucestershire	*E. Drinkwater*, 1 Vale Cottages, Wyck Hill, Stowe on the Wold, Cheltenham GL54 1HT
Guernsey BKA	*Ruth Collins*, Colombier House, Torteval, Guernsey, Channel Islands
Gwent	*R. Laxton*, 44 Beaumaris Drive, Llanyravon, Cwmbran, Gwent NP44 8JA
Hampshire	*Mary Dartnall*, 2 Harlyn Road, Southampton SO1 4NF
Herefordshire	*Reg Fennah*, Highfields, Wormbridge, Hereford
Hertfordshire	*Jim Crawford*, 14 Ridgeway, Radlett WD7 8PR
Huntingdonsh.	*Bob Evans*, 55 Boxworth Rd., Elsworth, Cambridge CB3 8JQ
Isle of Wight	*Mary Pain*, Rock Cottage, Sandy Lane, Blackwater, Newport PO30 3BS
Kendal & S. Westmorland	
	Maurice Bottomley, 21 Hillcrest Drive, Slackhead, Beetham, Milnthorpe LA7 7BB
Kent	*John Crosskey*, 56 Main Road, Biggin Hill, Westerham TN16 3DU
Leics. & Rutland	*Andrew Warren*, 86 Sandringham Ave, Belgrave LE4 7NR
Lincolnshire	*Mrs. Wilkinson*, 48 Brigsly Road, Waltham, Grimsby
London	*M.D.U. Granville*, Flat 1, 32 Oppindans Road, London NW3 5AG
Mole A. Club	*Esther Molyneux*, 120 Surbiton Hill Park, Surbiton, Surrey
Newcastle	*George Batey*, Rift Farm Cottage, Wylam NE41 8BL
Norfolk	*Paul Metcalfe*, Norfolk College of Agriculture, Easton, Norwich NR9 5DX
Northamptonsh.	*June Dennis*, 50 Station Road, Cogenhoe, Northampton
Nottinghamsh.	*Jennifer Cousins*, 148 Chesterfield Rd. South, Mansfield NG9 7AP
Shropshire	*Peter Woodcock*, 9 Buildwas Road, Wellington, Telford TF1 3NZ
Somerset	*Mike Milton*, Homestead, South Hill, Somerton TA11 7JQ
Suffolk	*Z. Ruskin*, 41 Bell Lane, Kesgrave, Ipswich, Suffolk IP5 7JH
Surrey	*D. Daniels*, Merrist Wood Ag. College, Worpleston, Guildford, Surrey
Warwickshire	*Brian Milward*, 13 Ash Drive, Kenilworth, Warwickshire
West Cornwall	*Andrew Reeve*, Vroe Farm, Mullion, Helston, Cornwall
Wiltshire	*Robin Chapman*, Hill Rising, Corsley, Warminster BA12 7PG
Worcester	*Harry Bradbear*, Horizons, Leigh, Worcester WR6 5LB

COUNCIL OF THE NATIONAL BEEKEEPING ASSOCIATIONS OF THE UNITED KINGDOM

Secretary: *Ken Ibbotson*
Brooklyns End
Murton
Co. Durham   SR7 9NR
☎ and **Fax:** (0191) 526 1893

Chairman: *D.J. Little,* Little Greengarth, 60 Deben Avenue, Martlesham Heath, Ipswich, Suffolk   IP5 7QP
Vice Chairman: *Scottish Beekeepers' Association nominee*
Treasurer: *J. Stoakley,* Drumlin, Craigerne Lane, Peebles   EH45 9HQ
☎ (01721) 200 97

CONBA was established in 1978 to promote the objects of the members: British, Scottish, Ulster and Welsh Beekeepers' Associations to consult and negotiate with local, national and international government authorities on their behalf, with a united voice.

The full Council meets twice a year, at the National Honey show in London and at the BBKA Convention at Stoneleigh. Observers from the Bee Farmers' Association, the Federation of Irish Beekeepers' Associations and the National Beekeeping Unit at Luddington attend the meetings. This provides an important forum for discussion of mutual beekeeping matters between national beekeeping organisations within the British Isles, government authorities and any other bodies.

CONBA is affiliated to the National Farmers' Union which enables it to represent its members on the COPA/COGECA Honey Working Party (HWP) in Brussels. CONBA therefore has a voice in discussions with the European Economic Community (EEC) and the European Parliament.

HWP meetings provide contacts with beekeeping organisations in mainland Europe. Legislation on honey, food regulations and bee diseases, originating in both the UK and the EEC, is monitored and representations made to both legislative bodies on behalf of Member Associations and of beekeepers generally.

DEVON APICULTURAL RESEARCH GROUP

Chairman: *Henry Morris*
4 Peryn Road
Tavistock
Devon   PL19 8LP

Secretary: *Bob Ogden*
Pennymoor Cottage
Pennymoor
Tiverton
Devon   EX16 8LJ

Publications Officer:
Ray King, 'Crossby', Peoples Park Road, Crediton, Devon   EX17 2DA
☎ (01363) 772007

DARG is a group of experienced beekeeper who seek solutions to current beekeeping problems by undertaking research in order to create and develop "best practices" and improved management techniques. The results of their work are frequently published as inexpensive booklets so that fellow beekeepers may benefit from the activities of the group.

Publications available are:
- *The Beeway Code* A common sense guide for beginners to help avoid problems with neighbours and produce a safe and peaceful apiary. (£1.80)
- *Seasonal Management* An easy to read no-nonsense booklet. Its many useful tips take you through the beekeeping year. (95p)
- *A Guide to Swarm Control* An easyily followed booklet covering the procedure of swarm control, with pocket logic charts to take in the apiary. (95p)
- *Living with Varroa Jacobsoni* DARG's best selling booklet, updated with the latest available control methods. (£1.80)
- *Practical Queen Rearing* All beekeepers should produce their own queens. This booklet covers many methods in a clear, concise manner, allowing you to choose the method that suits you best. (95p)
- *Selection of Apiary Sites* An invaluable guide, packed with tips on site selection, bee forage, access and security, safety hints and hazard spotting, providing shelter from the weather, moving bees, and much, much more. (£1.80)

All single copy prices include postage and packing. For bulk purchases, please contact the Publications Officer for details of discounted prices.

THE FEDERATION OF IRISH BEEKEEPERS' ASSOCIATIONS

Hon. Secretary: *Peter O'Reilly*
11 Our Lady's Place
Naas
Co. Kildare
☎ (045) 897568

OFFICERS

President: *Mr. Jim Ryan*, Inisfail, Kickham Street, Thurles, Co. Tipperary
☎ (0504) 22228

Vice President: *Mr. Michael O'Callaghan*, Say on Aran, Riverstown Bridge, Glanmire, Co. Cork ☎ (021) 822183

Life Vice Presidents:
Rev. Bro. H.I. Behan, Monkstown Park College, Dunlaoghaire, Co. Dublin
Mr. D.J. Deasy, 45 Waltham Terrace, Blackrock, Co. Dublin
Mr. J.J. Doran, St. Jude's, Mooncoin, via Waterford
Mr. M.I. Moynihan, 41 Caseyville, Dungarvan, Co. Waterford
Mr. P. O'Reilly, 11 Our Lady's Place, Naas, Co. Kildare
Mr. M.L. Woulfe, Railway House, Midleton, Co. Cork ☎ (021) 631011

Hon. Editor: *Eddie O'Sullivan*, St. Ives, Kilcrea Park Magazine Road, Cork
☎ (021) 542614

Hon. Manager: *Seamus Reddy*, 8 Tower View Park, Kildare ☎ (045) 21945

Hon. Treasurer: *Frances Kane*, Firmount, Clane, Co. Kildare

Education Convener:
Dr. Brendan Coughlan, Ard-na-Cloch, Corcullen, Moycullen, Co. Galway
☎ (091) 85211

Summer Course: *Michael Woulfe*, Railway House, Middleton
Co. Cork Convener: ☎ (021) 631011

PUBLICATIONS

- *Beekeeping in Ireland - A History* - *J.K. Watson*
  This book gives the history of the craft from time immemorial to the present. It is well bound, hard backed and excellently presented. There are 116 pages of valuable information and 53 pictures of prominent beekeepers past and present. Price £6.00
- *An Beachaire* - The Irish Beekeeper Monthly organ of FIBKA, subscription £10.00stg post free from The Manager. Readership of the Journal in Northern Ireland carries third party insurance up to £1,250,000 on any one claim, on payment of £4.00stg extra.

LIBRARY
The Library is owned and controlled by FIBKA. It contains very many valuable books ancient and modern, available to members for return postage only. The Librarian is Jim Ryan, Innisfail, Kickham Street, Thurles, Co. Tipperary.

CORRESPONDENCE COURSES
The Examinations Board has sponsored correspondence courses for candidates preparing for the Intermediate and Senior (Bee Masters) Examinations. Applications to John Cunningham, Main Street, Tramore, Co. Waterford.

EXAMINATIONS
The Board conducts five grades of examinations at the annual Summer Course at Gormanston College: Preliminary, Intermediate, Senior, Lecturer, Honey Judge. Preliminary Examinations are also held at provincial centres in May each year.

EDUCATION
Annual courses for beginners are run by Affiliated Associations and FIBKA holds an annual course at Gormanston College in July. The Course caters for 3 grades of students: beginners, intermediate and senior. Ten of the best Beekeeping Lecturers speak at these courses. The Grand Finale of the National Beekeeping Quiz is held on Thursday of the Course week. The winner is rewarded by the presentation of a Perpetual Trophy with prizes for runners up.

NATIONAL HONEY SHOW
This is held at Gormanston College in conjunction with the annual Beekeeping Course. The Schedule contains 29 Open Classes and 3 Confined Classes. Over 30 Challenge Cups and Trophies are presented for competition.

INSURANCE
The limit of indemnity of the public liability policy is £1,250,000 arising from one accident or a series of accidents. The policy extends to all registered affiliated members whose subscriptions are fully paid up on 31 December of any one year and whose names are entered in the FIBKA register held by the Treasurer. The Policy also extends to cover food poisoning.

ASSOCIATION SECRETARIES

Ashford	Mr. Pat O'Connell, Valentia, Miltown Lane North, Ashford, Co. Wicklow ☎ (0404) 40198
Banner	Mr. Kieran Byrnes, Gurteen, Tulla, Co. Clare ☎ (065) 35620
Bray	Ms. Wendy Bass, Berghoff, Glencree Road, Enniskerry, Co. Wicklow
Carbery	Mr. S. O'Brien, 71 North Street, Skibbereen, Co. Cork ☎ (028) 21152
Carlow	Mr. John Lennon, 31 Idrone Park, Tullow Road, Carlow
Co. Cavan	Mr. Charles Robinson, Milltown, Belturbet, Co. Cavan
Co. Cork	Mrs. Mary O'Riordan, Cooltubrid, Carigrohane, Co. Cork ☎ (021) 871098
Co. Dublin	Mr. Don Tarleton, "Carrig", Knocknacree road, Dalkey, Co. Dublin ☎ (01) 2858058
Co. Galway	Dr. Brendan Coughlan, Chemistry Dept., UCG, Galway ☎ (091) 85211
Co. Kerry	Mr. Neal Timlin, Clogherbrien, Tralee, Co. Kerry ☎ (066) 25724
Co. Limerick	Mr. John Donworth, Raheen, Ballyneety, Co. Limerick ☎ (061) 351551

Co. Longford	*Mrs. Brigit Koston*, Sunnyside House, Loughgowna, Co. Cavan ☎ (043) 83285
Co. Mayo	*Mrs. Mary Clarke*, Convent Terrace, Ballina, Co. Mayo
Co. Meath	*Mrs. Jane buckley*, Piercetown, Dunboyne, Co. Meath ☎ (01) 8255437
Co. Offaly	*Mrs. C. Rainsberry*, Briarly, Clonminch, Tullamore, Co. Offaly ☎ (0506) 51918
Co. Roscommon	*Mrs. Christina Waldron*, Athleague, Co. Roscommon
Co. Waterford	*Mrs. Claire Chevasse*, Cappagh House, Cappagh, Co. Waterford
Co. Westmeath	*Mr. Patrick McArdle*, V.S. Kilpatrick, Mullingar, Co. Westmeath ☎ (044) 42031
Co. Wexford	*Mr. Tom Quigley*, Ballyfarnogue, Screen, Enniscorthy, Co. Wexford ☎ (053) 37320
Dunamaise	*Mrs. Noreen Savage*, Derryguile, Mountmellick, Co. Laois ☎ (0502) 24331
Dunmanway	*Mr. M.L. O'Sullivan*, Ballyhalwick, Dunmanway, Co. Cork ☎ (023) 45257
East Cork	*Mr. C. Terry*, "Ait na Graine", Coolbay, Cloyne, Co. Cork
East Waterford	*Mr. N. Callanan*, "Naomh Josaf", 19 Manor Lawn, Waterford
Fingal	*Capt. O.J. McAlinden*, 8 Offington Drive, Sutton, Co. Dublin ☎ (01) 324718
Innishowen	*Ms. Maire Crumlish*, Churchtown, Carndonagh, Co. Donegal
Kilternan	*Ms. D. Smith*, 43 Elton Park, Sandycove, Co. Dublin ☎ (01) 2805676
Mid Kilkenny	*Mr. Jack Pass*, The Hollies, Gowran, Bennet's Bridge, Co. Kilkenny
New Ross	*Seamus Kennedy*, Churchtown, Fethard-on-Sea, New Ross, Co. Wexford
North Cork	*Ms. H. Noonan*, Coolroe, Fermoy, Co. Cork ☎ (025) 31496
North Kildare	*Mr. Sean Garrett*, 1608 Pairc Muire, Droichead Nua, Co. Kildare ☎ (045) 31260
N. Monahgan	*Mr. Francis McQuinlan*, 43 Belgium Park, Monaghan
N. Tipperary	*Mr. Tony Casey*, Clongour, Thurles, Co. Tipperary ☎ (0504) 21963
Roundwood	*Mrs. M. O'Byrne*, "Carrig View", Moneystown South, Roundwood, Co. Wicklow ☎ (0404) 45209
South Kildare	*Bro. H.I. Behan*, CBS, Portarlington, Co. Laois
S. Kilkenny	*Mrs. I.E. Jones*, "Moytura", Knock House, Upper Gracedieu, Co. Waterford ☎ (051) 73218
S. Tipperary	*Mr. Dennis Ryan*, Mylerstown, Clonmel, Co. Tipperary
South Wicklow	*Tom McBride*, Gleendra, Ballynattin, Arklow, Co. Wicklow ☎ (0402) 39941
S.W. Cork	*Mr. John Bryan*, Currarane, Kilbrittan, Co. Cork ☎ (023) 49625
West Cork	*Mr. Urs Feuer*, Dromore, Bantry, Co. Cork
Killorglin	*Mr. Declan Crowley*, Gurrane, Killorglin, Co. Kerry ☎ (066) 61710
N.W. Donegal	*Michael Gallagher*, Pluck, Letterkenny, Co. Donegal ☎ (074) 57114

INTERNATIONAL BEE RESEARCH ASSOCIATION

Correspondence to:
> Director
> International Bee Research Association
> 18 North Road
> Cardiff CF1 3DY
> ☎ (01222) 372409 **Fax:** (01222) 665522
> **E-mail:** ibra@cardiff.ac.uk

IBRA - the International Bee Research Association - is the world's premier beekeeping information service. It is a charity devoted to advancing apicultural education and science worldwide.

Since its formation in 1949 IBRA has worked to serve the information needs of beekeepers, scientists, advisers, teachers and many other groups. Technically-minded beekeepers in the UK and in other countries find IBRA's journals and library services uniquely valuable.

PUBLICATIONS
IBRA's three journals are highly respected among beekeepers and bee scientists.

- *Bee World* provides beekeepers and scientists with authoritative articles and reviews on topical issues.
- *Journal of Apicultural Research* is an award-winning English-language journal containing up-to-date original research articles.
- *Apicultural Abstracts* reviews scientific and practical literature on apiculture and bee pollination of crops. It provides a way of scanning the world's libraries from your own home.

IBRA also publishes authoritative texts, pamphlets, educational aids, bibliographies and multilingual dictionaries.

INFORMATION ON-LINE
B.mail is a monthly newsletter produced by IBRA and available free of charge by direct subscription and on several newsgroups. It's filled with news and short articles on topical subjects.

IBRA also provides a free information service on the World Wide Web. Visit our pages at:
> http://www/cardiff.ac.uk/ibra/index.html

LIBRARY

The IBRA library holds a wealth of information unequalled elsewhere in the world. It includes 35,000 books and reprints, along with more than 3,000 volumes of beekeeping journals and theses, reports and bibliographies. The library is continually being added to with contributions from many countries.

Members can borrow books and people from all over the world have access to this library through our information search and reprint services. Members can use BeeSearch - our trained librarian can carry out computer searches to find information on any subject and can provide copies of specific papers. Slides on subjects related to beekeeping are also available for loan.

MEETINGS

Another important information service provided by IBRA is the organization of conferences and other meetings. IBRA arranges technical meetings for UK beekeepers and is also involved in co-sponsoring conferences with other specialist societies.

We also convene a unique series of international conferences on apiculture in tropical climates, spanning 20 years so far. These conferences are a forum for discussing practical issues concerning beekeeping in the tropics. The proceedings are published by IBRA and have become valued textbooks.

PUBLICATION SALES

The publication sales unit at IBRA sells books, audio visual material and other items to readers all over the world. This service gives everyone easy access to beekeeping literature, and every time you buy something from our bookshop, your money is helping to sustain our work.

ADVISORY SERVICES

IBRA has considerable expertise in consultancy projects and has worked on contract for international agencies, governments and non-governmental organizations. Our staff have extensive international experience, having carried out consultancy missions or study visits to countries in all continents.

MEMBERSHIP

The key to IBRA's work is support from members. Their subscriptions provide a large part of IBRA's income and their contributions of information and voluntary promotional work greatly help the association's activities. Members receive a free subscription to *Bee World*. They are welcome to attend, speak and vote at our annual general meeting and nominate members to serve on IBRA's council. We encourage members to visit IBRA headquarters in Cardiff and use the library.

INSTITUTE OF APICULTURE

Secretary:	Institute of Apiculture
	Bee Research Unit
	School of Biology
	University of Wales
	Cardiff CF1 3TL
	☎ (01222) 874312 **Fax:** (01222) 874305

COUNCIL:

Chairman:	*Prof. R.S. Pickard*
Treasurer:	*Dr. Nicola Bradbear*

Prof. J.B. Free,,Mr. R.E. Gove, Prof. L.A.F. Heath, Mr. G.W. Knights, Mr. I.S. McLean, Mr. R.W. Sawyer, Mr. L.L. Thorne, Mr. A.C. Waring

The principal objective of the Institute of Apiculture is to establish, somewhere in the United Kingdom, a permanent and independent experimental research centre for the study of bees and beekeeping. The IOA has developed within the Bee Research Unit of the University of Wales at Cardiff, which is the largest centre of bee research in the UK and has attracted many donations from beekeepers and their associations since its foundation in 1977.

Members of the Institute receive the following benefits:

- Free copies of "*HONEYBEE*", the IOA house journal
- Reduced attendance fees at the Bee Research Unit's annual training courses. These include:
    - a) beekeeping
    - b) pathology
    - c) queen propogation
    - d) microscopic examination of honey
    - e) instrumental insemination
- Access to information on all aspects of honeybee biology and apiculture.
- Reduced fees for any services provided by the Bee Research Unit for the apicultural industry. These include the identification of pollen in honey, Acarapis and Nosema.

At Cardiff students can investigate any aspect of honeybee biology or any practical problem in apiculture at either Final Year BSc., Diploma, M.Phil. or PhD level. They have access to the School of Biology's three electron micro-scopes, analytical laboratories, computing services and technical expertise in most branches of the biological sciences. Fieldwork is undertaken at local out-apiaries in Wales and international projects are underway in Mexico, Pakistan and Portugal. Students are currently recruited from over 50 different countries throughout the world and a practical contribution is made to the establishment of low cost, low technology apicultural industries.

NATIONAL PERIODICALS

American Bee Journal - Agents: *Steele & Brodie*, Newport on Tay, Wormit, Fife.

Apiacta -   International Beekeeping in English, French, German and Spanish by the International Fed. of BKAs.
E.H. Thorne (Beehives) Ltd., Beehive Works, Wragby, Lincoln   LN3 5LA.

Australasian Beekeeper - Monthly.
Sample from: *PMB*, 19 Maitland, NSW 2320

Australian Bee Journal - P.O. Box 365, Emerald, Victoria 3782, Australia.

Bee Biz -   A new magazine for commercial beekepers. Published 3 times per year. Subscriptions £12 p.a. from: *Northern Bee Books*, Scout Bottom Farm, Mytholmroyd, Hebden Bridge, W. Yorkshire   HX7 5JS

Bee Craft -   Official monthly journal of the British BKA. Subscriptions and enquiries to: *Secretary*, 15 West Way, Copthorne Bank, Crawley, Sussex.

Beekeepers News - Quarterly by:
E.H. Thorne (Beehives) Ltd., Beehive Works, Wragby, Lincoln   LN3 5LA

The Beekeepers Quarterly - Companion to the Beekeepers Annual.
Subscriptions £12 p.a. from: *Northern Bee Books*, Scout Bottom Farm, Mytholmroyd, Hebden Bridge, W. Yorkshire   HX7 5JS

Beeline -   P.O. Box 192, Borrowdale, Harare, Zimbabwe.

British Bee Journal - Monthly.
Subscriptions and enquiries to: *British Bee Publications Ltd.*, 46 Queen Street, Geddington, Kettering   NN14 1AZ

Canadian Bee Journal - 47 Black Knight Road, St. Catherine's, Ontario, Canada

Canadian Beekeeping - P.O. Box 128, Orono, Ontario, Canada   LOB 1MO

Gleanings in Bee Culture - US monthly. *E.H. Thorne Ltd.*, Wragby   LN3 5LA

Indian Bee Journal - In English.
1325 Sadashiv Peth, Poona 411 8030, India.

International Bee Research Association - samples:
Bee World 50p; *Journal of Apicultural Research* 50p; *Apicultural Abstracts* 75p. 18 North Road, Cardiff

Irish Beekeeper - Monthly.
Editor: *Eddie O'Sullivan*, St. Ives, Kilcrea Park, Magazine Road, Cork

New Zealand Beekeeper - Published quarterly for National BKA of New Zealand.
Subscriptions: Dept. Box 4048, Wellington NZ.

The Scottish Beekeeper - Magazine of the Scottish BKA. Membership terms from:
D Blair, 44 Dalhousie Road, Kilbarchan   PA10 2AT. Sample copy 22p.

South African Bee Journal - Bi-monthly.
P.O. Box 41, Modderfontein, 1645, RSA.

The Speedy Bee - Monthly US newspaper, *E.H. Thorne Ltd.*, Wragby, Lincoln   LN3 5LA

Welsh Bulletin/Newyddlen - quarterly magazine of the Welsh BKA.
Editor/Golygydd: *Miss A.L. Jones*, 13 Maes Meugan, Llanrhydd, Rhuthun LL15 1YH

THE NATIONAL DIPLOMA IN BEEKEEPING

Hon. Secretary: *Reg Gove* NDB FRES
Westcott
Gerway Lane,
Ottery St. Mary   EX11 1PW

Chairman: *Geoff Hopkinson* NDB,
39 Old Croft Road, Walton-On-The-Hill, Stafford   ST17 0NJ

The Examination Board for the National Diploma in Beekeeping was set up in 1954 to meet a need for a beekeeping qualification above the level of the highest certificate awarded by Beekeeping Associations and approaching degree level.

The need was manifest as an appropriate qualification for a County Beekeeping Lecturer or a specialist appointment requiring a high level of academic and practical ability in beekeeping. It is the highest beekeeping qualification recognised in the United Kingdom and those who obtain the Diploma can feel justly proud.

The Board consists of representatives from a wide range of organisations and from Government Departments and together form an impressive amalgam of expert knowledge in beekeeping and education. Although the National Beekeeping Associations are represented on the Board it is entirely independent of them.

Normally the highest certificate of one of the National Associations is a necessary criterion for eligibility to take the Examination for the Diploma which is held in alternate years and extends over two days. The Examination consists of two written papers of three hours each and viva-voce, plus practical tests conducted by at least four Examiners appointed by the Board.

The Board also organises an Advanced Beekeeping Course covering certain parts of the syllabus, which can be difficult to study at home. For many years this was held at the Hampshire College of Agriculture, Sparsholt but with the demise of the Bee Department at that College the Board is cooperating with the Gloucestershire College of Agriculture & Horticulture, Hartpury House, Gloucester, with the aim of running similar courses at Hartpury in future years. The course lasts a working week and covers a wide field including laboratory techniques, the recognition of the main brood diseases in the comb, microscopy in relation to beekeeping, dissection of the honey bee, general laboratory techniques and the recognition of pollens and bee plants. Visits are made to commercial beekeepers and to large scale honey packers. The outside lecturers are acknowledged leaders in their fields.

For further details write, enclosing s.a.e., to the Secretary.

List of all those who have gained the National Diploma in Beekeeping:

- Matthew Allan
- Harry Allen
- Harrison Ashforth
- John Ashton
- Dianne Askquith-Ellis
- John Atkinson
- Miss. E. Avey
- Brig. H. Bell
- P.W. Brooke
- Mrs. Rosina Clark
- Charles Collins
- Gerry Collins
- Tom Collins
- Mrs. C. Davis
- John Cowan
- S.J. Cox
- Jim Crundwell
- Celia Davis
- Ken Stevens
- J. Swarbrick
- Margaret Thomas
- John Walker
- Adrian Waring
- Brian Welch
- T. Wilbraham

- Reg Gove
- Eric Greenwood
- A.R.W. Griffin
- Robert Hammond
- C.A. Harwood
- Leslie Hender
- Alf Hebden
- Ted Hooper
- Geoff Hopkinson
- G. Howatson
- Geoff Ingold
- George Jenner
- C.F. Jesson
- A.C. Kessel
- W.E. Large
- G.W. Lumsden
- Henry Luxton
- Alec S.C. Deans
- Clive de Bruyn
- A.P. Draycott
- M. Feeley
- Barry Fletcher
- David Frimston
- Oonagh Gabriel
- George Gill

- G. N'tonga
- Peter Oldrieve
- Gillian Partridge
- E.H. Pee
- L.E. Perera
- E.R. Poole
- Bill Reynolds
- Fred Richards
- E. Roberts
- Arthur Rolt
- Jeff Rounce
- J. Ryding
- J.H. Savage
- A. Sims
- Dr. F.G. Smith
- George Smith
- J.H.F. Smith
- A.S. McClymont
- Ian McLean
- J.L. MacGregor
- Ian Maxwell
- Paul Metcalf
- J. Mills
- Bernhard Möbus

THE NATIONAL HONEY SHOW

Hon. General Secretary:
> *Rev. H.F. Capener*
> 1 Baldric Road
> Folkestone   CT20 2NR

President: *J.W. Holt*
Hon. Treasurer/Membership Sec.:
> *W.W. Jones*, 2 Vicarage Gardens, Bradwell, Milton Keynes   MK13 9AY

Chairman: *W.S. Mundy*
Vice Chairman: *Mrs. J. Purcell*

At the National Honey Show each year, beekeepers from all over the British Isles, the Republic of Ireland, the Channel Islands and the West Indies join together, not only in the friendly competition of the show-bench, but also to take advantage of the programme of superb lectures of the BBKA Convention.

All those who attended the 1993 and 1994 Shows will vouch for the excellence of Kensington Town Hall, where the facilities for the exhibition, the lectures and for car-parking are all that could be asked for.

The National Honey Show is organised entirely by voluntary helpers, who will always gladly welcome any offers of assistance. For further information, please write to the Hon. General Secretary, whose name and address appears above.

QUEEN MARY WESTFIELD COLLEGE LONDON UNIVERSITY

Contact: *Dr. Lesley Goodman*
School of Biological Sciences
Queen Mary College
Mile End Road
London   E1 4NS
☎ (0171) 775 3296

Team: *Mr. C.R. Masri, Mr. K. Pell, Mr. D. Mapp*

Research into bee behaviour and physiology has been carried out at QMW by a group under the leadership of Dr. Lesley Goodman for the past fifteen years. The overall objective of the work undertaken at QMW is to increase our understanding of bee behaviour by the study of bee senses and the bee nervous system. The QMW group is particularly concerned with the visual system of the bee and the neural mechanisms underlying the bee's use of visual cues in navigation, orientation and course correction during flight. Recent work has revealed the system of neurons which underlie the bees' ability to correct for deviations in the yawing, rolling and pitching planes during flight. We are now investigating the integration of visual and olfactory information in the foraging bee.

Past work includes elucidation of the wiring of the **'mushroom bodies'**, important integration areas within the brain and, leading on from this, the group is beginning to look at the integration of visual and olfactory information in the brain. This should help us to further understand the neural mechanisms underlying the selection and subsequent relocation of forage. Other work in progress includes an examination of microclimate control in the hive throughout the year. Two experimental hives have been built in which the temperature distribution in every frame and the humidity and light intensity are automatically monitored at ten minute intervals throughout the year. In this way a detailed picture of the microclimate of the hive has been obtained over an extended period. Using one hive as a control, the effects of novel hive materials and minor, cheap manipulations of temperature and light on the increase of brood and build up of foraging capacity in the spring are being investigated.

ROTHAMSTED EXPERIMENTAL STATION

Institute of Arable Crops Research
Entomology and Nematology Department
Rothamsted Experimental Station
Harpenden
Hertfordshire   AL5 2JQ
☎ (01582) 763133

Bee Project Leaders:
    Prof. Ingrid Williams - pollination ecology
    Ms. Brenda Ball - honeybee pathology
Scientific Staff: *Mr. Mark Allen, Dr. Juliet Osborne, Dr. Guy Poppy,*
    Ms. Jacqueline Simpkins
Apiculturist: *Mr. Norman Carreck*
RI Fellow: *Dr. Dejair Message*
PhD Student: *Mr. Colin Denholm*

Rothamsted is the oldest laboratory in the world devoted to agricultural research, having started in 1843, and has a total staff of around 500. Research on bees at Rothamsted has been continuous since 1923 and a number of eminent bee scientists, among them Dr. Bailey, Dr. Butler, Dr. Free, Dr. Ribbands, Dr. Simpson and Dr. Stevenson were pioneers in their particular research areas.

The expertise at Rothamsted in honey bee pathology and pollination ecology is unparalleled and the modern laboratories with a range of specialised equipment facilitate research on diverse topics. Experienced apiary staff maintain c.100 honey bee colonies at different locations and the bee field station houses a behaviour laboratory, observation hive room and bee flight room. Facilities for field crop studies are available on site and experimental design and analysis are backed up by good statistical support.

RESEARCH PROGRAMME
Prof. Ingrid Williams leads a group investigating the interaction between bees, crops and the environment and the efficient management of native bees for crop pollination. Pollination studies focus on oil-seed rape, field beans, sunflowers, linseed, lupins and white clover. Research into the foraging behaviour of bees is aimed at modelling the numbers of pollinators needed for optimal fertilization, determining its relationships with pollen and nectar production and weather, investigating pollen flow through and from crops and the interactions between different bee species. The spatial and temporal foraging behaviour of honey bees and bumble bees within agricultural areas is being compared and the potential value of good bee forage plants to enhance arable environment for bees by providing successional forage is also being

investigated. A new programme on the role of semiochemicals in regulating foraging behaviour and bumble bee colony development has been initiated. The work aims to help ensure sustainable production of entomophilous fruit and seed crops while minimizing environmental damage.

Research in the honey bee pathology section, led by Ms. Brenda Ball, has included studies on the properties, natural histories and means of control of protozoan, mite, fungal, bacterial and viral infections of bees. Current research aims to determine whether economic damage to varroa-infested honey bee colonies in the UK is associated only with parasitization of adult bees and pupae by the mite or whether significant losses are due to elevated levels of other pathogens. Investigation of the factors affecting disease epidemiology in infested colonies and the role of the mite as an activator and vector of infections should provide options for the development of novel management or control strategies to reduce the damaging effects of mite infestation. A new programme to determine the mechanism of virus induction in bees parasitized by varroa has been initiated.

A one year Rothamsted International Fellowship has enabled Dr. Dejair Message of the Federal University of Viçosa to investigate the cause of a serious honey bee brood disease in Brazil. Such observations on pathogens that appear to be exotic will enable a rational policy on the importation of honey bees to be maintained and equip British beekeepers with the knowledge necessary to reduce the impact of such infections should they become established.

INFORMATION EXCHANGE
Expertise in bee research is drawn upon by scientific colleagues world-wide and there are research links with institutes and universities in this country and abroad. Working visits are arranged annually for overseas scientists wishing to gain experience or specialist skills in honeybee research and they make a valued contribution to established programmes.

Research findings are published in scientific journals but popular articles are also written for the beekeeping and agricultural press. Staff members serve on both national and international committees on diverse aspects of apiculture and lectures are presented to national and local beekeeping associations.

FUNDING
Rothamsted now forms part of the Institute of Arable Crops Research and receives funds for research from the Biotechnology and Biological Sciences Research Council and through commissions and contracts from the Ministry of Agriculture, Fisheries and Food, from Levy boards, commercial and other organisations.

THE SCOTTISH BEEKEEPERS' ASSOCIATION

General Secretary:

> *Mrs. Una A. Robertson*
> North Trinity House
> 114 Trinity Road
> Edinburgh   EH5 3JZ
> ☎ (0131) 552–5341

Hon. President: *The Rt. Hon. Earl of Mansfield* D.L., J.P., Scone Palace, Perth   PH2 6BE
Hon. Vice Pres.: *A.S.C. Deans*, 21 Port View, Port Seton, East Lothian   EH32 0TX
A.J. Lilburn, Mains of Coull, Aboyne   AB3 4TS
G.C. Smith, 12/27 Ethel Terrace, Edinburgh   EH10 5NA
Hon. Librarian: *Mrs. Margaret M. Sharp*, City Librarian, City Library, George IV Bridge, Edinburgh
Hon. Legal Adviser:

> *George Mathers & Co.*, 23 Adelphi, Aberdeen   ☎ (01224) 588599

EXECUTIVE COMMITTEE

President: *William A. MacKenzie*, 9 Glenhome Avenue, Dyce, Aberdeen   AB2 0FF
☎ (01224) 722598
Vice President: *Ian Craig*, 30 Burnside Avenue, Brookfield, Johnstone, Renfrewshire
PA5 8UT   ☎ (01505) 322684
Imm. Past Pres.: *Iain F. Steven*, 2 Viewforth Road, South Queensferry   EH30 9MF
General Sec.: *Mrs. Una A. Robertson*, North Trinity House, 114 Trinity Road, Edinburgh
EH5 3JZ   ☎ (0131) 552 5341
Treasurer: *Douglas J. Gillespie*, 26 Union Road, Inverness   IV2 3JY   ☎ (01463) 232424
Editor, Scottish Beekeeper:

> *Mrs. E. Morna Stoakley*, Drumlin, Craigerne Lane, Peebles   EH45 9HQ
> ☎ (01721) 720097

CONVENERS OF STANDING COMMITTEES

Membership: *Fraser Sim*, 27 Moss Road, Tain, Ross-shire   IV19 1HH   ☎ (01862) 892351
Compensation & Insurance:

> *C. Gordon Stewart*, 113 Cumbernauld Road, Stepps, Glasgow   G33 6EU
> ☎ (0141) 779–1899

Education: *Ian Craig*, 30 Burnside Avenue, Brookfield, Johnstone, Renfrewshire
PA5 8UT   ☎ (01505) 322684
Publicity: *David B.N. Blair*, 44 Dalhousie Road, Kilbarchan, Renfrewshire   PA10 2AT
☎ (01505) 702680
Shows: *Peter C. Aird*, Craigview Cottage, West Lynn, Dalry, Ayrshire   KA24 4LJ
☎ (0129) 483 2081

Library:	*Ian Will*, 25 St. James View, Penicuik, Midlothian EH26 9DZ ☎ (01968) 677015
Markets:	*L.M. Webster*, Birchlea, Rothiemay, Huntly, Aberdeenshire AB54 5LN ☎ (01466) 771351

Area Representatives:

North	*M.D. Canham*, Whinhill Farmhouse, By Cawdor, Nairn IV12 5RF ☎ (01667) 404314
East	*T. McGravie*, Birnam, 12 Foulis Crescent, Juniper Green, Edinburgh EH14 5BN ☎ (0131) 453 4004
West	*Mrs. E. Ann Fisher*, Ashcraig Cottage, Shore Road, Skelmorlie, Ayrshire PA17 5HB ☎ (01475) 520336
Hon. Auditor:	*L. Stan Coxon*, 10 Eriskay Road, Inverness IV2 3LX ☎ (01463) 240588

OFFICERS

Spray Liaison Co-ordinator and Varroa Officer:

John T. Stoakley, Drumlin, Craigerne Lane, Peebles EH45 9HQ
☎ (01721) 720097

AVA Officer: *Alistair J. Lilburn*, Mains of Coull, Aboyne AB3 4TS ☎ (01339) 881 452

AIMS OF THE ASSOCIATION

To bring together all those interested in beekeeping to the benefit of horticulture and agriculture by providing helpful educational facilities viz:
- to publish a monthly magazine
- to maintain the Moir Library in Edinburgh
- to conduct examinations in the art of beekeeping
- to provide insurance and a compensation scheme for members

S.B.A. LECTURERS

* Addresses in SBA Honey Judges List
All those listed claim expenses
All speakers accompany talks with visual aids

NAME	TEL. NO.	SUBJECT
* R.G. Brown	(01387) 52773	General, Honey Preparation
* I. Craig	(01505) 322684	General
* A.B. Ferguson	(01461) 5322	General
* A.J. Lilburn	(01339) 881452	Beekeeping Abroad, Beekeeping for Beginners
* G.C. Smith NDB	(0131) 4475332	General
* J.H.F. Smith NDB	(01292) 442838	General
T. Smith Standalone Palnure, Kirkcudbrightshire	(01671) 2276	Beekeeping in the Sultanate of Oman, Apis Florea
A. McL. Stirrat 27 Marlepark, Ayr	(01292) 42833	General
* Mrs. M Stoakley } * J. Stoakley }	(01721) 720097	Various, including Beeswax, Honey Sources, Pheromones, Nuclei.
* W.B. Taylor	(01569) 740375	General

EDUCATION
The SBA arranges courses and awards certificates to successful candidates in the Scottish Basic Beemaster, Expert Beemaster and Honey Judge Examinations. It also actively promotes beekeeping by informing the public, especially the young, about bees and their benefits to the environment.

INSURANCE AND THE COMPENSATION SCHEME:
All members of the SBA have insurance against Public Liability. The SBA Compensation Scheme is restricted to be colonies located in Scotland and allocates part-replacement value for damage by vandalism, fire, theft and certain brood diseases.

LIBRARY
The SBA Moir Library in Edinburgh has one of the world's finest collection of beekeeping books. A library card is issued annually to every member who can borrow books at the cost of return postage only. Details may be obtained from the Library Convener.

MARKETS
Advice is given on all aspects of marketing honey products at appropriate times. Suggested bulk, wholesale and retail prices are notified in the magazine.

PUBLICATIONS
- *The Scottish Beekeeper* is published monthly and sent post free as part of the annual membership fee of £10 payable to the Membership Convener.
- *Introduction to Bees and Beekeeping* is £2.00 plus 30p postage and may be obtained from the Publicity Convener.

PUBLICITY
Members can purchase the association tie, lapel badge, car sticker etc. Details may be obtained from the Publicity Convener.

SHOWS
Three major annual honey shows are held in Scotland. They are at the Royal Highland Show, Ingliston, Edinburgh in June, the Scottish National Honey Show at the Ayr Flower Show in August and the Honey Show at the Dundee Flower Show in September. Details may be obtained from the Shows Convener.

MEMBER ASSOCIATIONS AND THEIR SECRETARIES

Ayr	R. Cunningham, 2A Townend Terrace, Symington KA13 5RT ☎ (01563) 830542
Border	Berwickshire and Kelso have amalgamated and are now Border BKA. Secretary: A.F. Mitchell, 22 Parkside, Coldstream TD12 4DY ☎ (01890) 882683
Bute	I. Chisholm, North Lodge, Ascog, Isle of Bute PA20 ☎ (01700) 504658
Caddonfoot	R. Kinsman-Blake, Kiln Cottage, Lilliesleaf Pottery, Main Street, Lilliesleaf, Melrose TD6 7JD ☎ (01835) 870432
Clyde Area	Dr. D. Christison, Lea Rig, 21 Eastcote Avenue, Glasgow G14 9LQ ☎ (0141) 959 5376
Cowal	Mrs. L.B. Pendreich, Westering, George Street, Hunters Quay, Dunoon PA23 8JU ☎ (01369) 3193
Dingwall	W.A. Hill, Fairfield, Orrui Budge, Muir of Ord, Ross-shire IV6 7UL ☎ (01997) 433 377
Dunfermline & West Fife	J. Tout, 13 Middlebank Holdings, By Dumfermline KY11 5QN ☎ (01383) 415534

East of Scotland *Mrs. H. Kinnes*, 3 Holly Road, Broughty Ferry   DD5 2LZ   ☎ (01382) 477762

East Lothian *M.J. Hill*, 3 Castlemains Cottages, Dirleton, North Berwick   EH39 5EG
☎ (0162085) 395

Easter Ross *Mrs. P. Douglas-Menzies*, Mounteagle, Fearn, Ross-shire   IV20 1RP
☎ (01862) 832213

Eastwood *K. Cowle*, 15 Wickham Avenue, Crookfur, Newton Mearns, Glasgow
G77 6AU   ☎ (0141) 639–6929

Edinburgh *A. Halkett*, 8 Pendreich Grove, Bonnyrigg, Midlothian   EH19 2EH
& Midlothian ☎ (0131) 663 7480

Fife *Mrs. A.E. McCallum*, Southfield, Pitscottie, Fife   KY15 5TX   ☎ (01334) 828521

Fortingall *R.M.R. Sturrock*, Rose Cottage, Taybridge Road, Aberfeldy   PH15 2BH
☎ (01887) 20533

Freuchie *Mrs. D.M. Stafford*, 2 Malt Loan, Newton of Falkland, via Cupar, Fife
KY7 7RZ   ☎ (01334) 857470

Glasgow *C.E. Irwin*, 55 Lindsaybeg Road, Chryston, Glasgow   G69 9DW
☎ (0141) 779 1333

Helensburgh *B.A. Bellamy*, Portkil House, Kilcreggan, Dunbartonshire   G84 0LF
☎ (01436) 842692

Inverness *R. Buckle*, Am Fasgadh, Tomatin, Inverness-shire   ☎ (01808) 511448

Kelvin Valley *D. Christison*, 21 Eastcote Avenue, Glasgow   G14 9LQ   ☎ (0141) 959 5376

Kilbarchan *I. Craig*, 30 Burnside Ave., Brookfield, Johnstone   PA5 8UT   ☎ (01505) 322684

Kilmarnock & Irvine
R.N.H. Skilling, 9 Dick Road, Kilmarnock   KA1 3AP   ☎ (01563) 21625

Kirriemuir *D.G. Norrie*, 14 Strathmore Ave., Kirriemuir   DD8 4DJ   ☎ (01575) 572194

Largs *Mrs. Ann Fisher*, Ashcraig Cottage, Shore Road, Skelmorlie   KA17 5HB
☎ (01475) 520366

Lochaber *P.J. Browne*, The Rowan Tree, Gairlochy, Spean Bridge, Inverness-shire
PH34 4EQ   ☎ (01397) 712730

Moray *Mrs. B. Vidler*, Garden Cottage, Moy, By Forres   IV36 0SP   ☎ (01309) 673888

Mull *Mrs. S. Barnard*, Viewmount, Tobermory, Isle of Mull   PA75 6PG
☎ (01688) 2008

Nairn *T.W. Rowe*, Post House, Littlemill, Nairn   IV12 5QL   ☎ (01667) 453292

Olrig *R. Inglis*, Roadside, Skirza, Freswick, Wick, Caithness   KW1 4XX
☎ (01955) 81260

Paisley *R.C. Adamson*, 25 Richmond Ave., Clarkston, Glasgow   G76 7JL
☎ (0141) 638–0752

Peeblesshire *Mrs. A. Hick*, Carcaut, Heriot, Midlothian   EH38 5YE   ☎ (01875) 835227

Perthshire *J. Shovlin*, 'Invercarse', 4 Glebe Terrace, Perth   PH2 7AG   ☎ (01738) 27965

S. of Scotland *R.G. Brown*, 6 Richmond Avenue, Dumfries   DG2 7JS   ☎ (01387) 52773

Stirling & Dunblane
T. Harley, 60 Westerlea Drive, Bridge of Allan, Stirling   FK9 4DQ
☎ (01786) 832558

Sutherland *Mrs. D.R. Royce*, West Garty Brae, 57 Culgower, Loth, Helmsdale,
Sutherland   KW8 6HP   ☎ (01431) 821535

West'n Galloway *L.N. Robertson*, Orchardton House, Acre Place, Wigtown, Newton Stewart
DG8 9DU   ☎ (01988) 402208

West Linton *D. Stokes*, 100 Main Street, Roslin, Midlothian   EH25 9LT   ☎ (0131) 440 3477

Aberdeen *S.C. Rae*, 2 Hartington Road, Aberdeen   AB1 6YA   (not affiliated)

Oban & District *M. Dobson*, "Whinfield", Glenmore Road, Oban   PH43 4NB

SBA HONEY JUDGES

M.J. Allan	41 George Street, Eastleigh, Hants. S050 9BT ☎ (01703) 617969
D.B.N. Blair	44 Dalhousie Road, Kilbarchan PA10 2AT ☎ (01505) 702680
G.H. Braithwaite	Glencoe Cottage, Dundee Road, Forfar DD8 1XD ☎ (01307) 462206
B.S. Brookes	Borelick, Trochry, Dunkeld PH8 0BX ☎ (01350) 723222
R.G. Brown	6 Richmond Avenue, Dumfries DG2 7JS ☎ (01387) 52773 or (01556) 690259
P.J. Browne	The Rowan Tree, Gairlochy, Spean Bridge, Inverness-shire PH34 4EQ ☎ (01397) 712730
M. Canham	Whinhill Farm House, by Cawdor, Nairn IV12 5RF ☎ (01667) 404314
I. Craig	30 Burnside Avenue, Brookfield, Johnstone, Renfrewshire PA5 8UT ☎ (01505) 322684
J. Culbert	The Shieling, Stanley, Perthshire PH1 4NF
D.H. Daniels	Twin Oaks, Nightingale Road, Ash, Aldershot, Hants
G.S. Duncan	36 Seafield Drive, Ayr
A.B. Ferguson	Firparkneuk, Kirtlebridge, Lockerbie DG11 3LZ ☎ (01461) 5322
W.J. Foubister	Westfield, Alford, Aberdeenshire AB33 8QA ☎ (01975) 562025
D.L. Hutchinson	4 Beach Road, Tynemouth NE30 2NT
W.J. Jones	Abalon, Llanfrothen, Grognedd LL48 6LJ
A.J. Lilburn	Mains of Coull, Aboyne AB3 4TS ☎ (01339) 881452
A.E. MacArthur	Melbourne House, Regent Street, Dalmuir, by Glasgow ☎ (0141) 952 1234
W.A. MacKenzie	9 Glenhome Avenue, Dyce, Aberdeen AB2 0FF ☎ (01224) 722598
H. Mclean	11 Stewart Crescent, Currie, Edinburgh EH14 5SE
I.C. Maxwell	Scottish Agricultural College, Auchincruive, Ayr KA6 5HW ☎ (01292) 520331
Mrs. A. Middleditch	11 North Lane, Norham, Berwick-upon-Tweed
B. Möbus NDB	5 Rue Fontfrede, Domaine des Alberes, Laroque des Alberes, F-66740 St. Genis des Fontaines, Roussillon, France
A. Murray	17 Victoria Crescent, Clarkston, Glasgow G76 ☎ (0141) 644 1714
Mrs. J. Purcell	275 Bath Road, Hounslow TW3 3DA
S.C. Rae	2 Hartington Road, Aberdeen AB1 6YA ☎ (01224) 322953
J.M. Robinson	Meadowbank, New Street, Kilbarchan, Renfrewshire PA10 2LN ☎ (015057) 2609
A.M. Ross	The Anchorage, 12 Low Town, Collieston, Aberdeenshire
J.T.W. Scruby	Pilgrim's Ridge, Markway, Godalming, Surrey GU7 2BW
R.N.H. Skilling	9 Dick Road, Kilmarnock KA1 3AP ☎ (01563) 21625
G.C. Smith	12/27 Ethel Terrace, Edinburgh EH10 5NA ☎ (0131) 447 5332
J.H.F. Smith	23 Taybank Drive, Ayr KA7 4RL ☎ (01292) 442838
Mrs. M. Stoakley	Drumlin, Craigerne Lane, Peebles EH45 9HQ ☎ (01721) 720097
J.T. Stoakley	Drumlin, Craigerne Lane, Peebles EH45 9HQ ☎ (01721) 720097
W.B. Taylor	West Newbigging Cottage, Glenbervie Road, Drumlithie, Stonehaven AB3 2YA ☎ (01569) 740375
L.M. Webster	Birchlea, Rothiemay, Huntly, Aberdeenshire AB54 5LN ☎ (01466) 86351
C. Weightman	Shilford, Stocksfield, Northumberland NE43 4HW ☎ (01661) 842824

ULSTER BEEKEEPERS ASSOCIATION

Secretary:	*Charles Nicholson* 57 Liberty Road Carrickfergus Co. Antrim BT38 9DJ ☎ (0960) 362998

President: *Jim Holland*
Chairman: *Duncan Saunders*
Vice Chairman: *Huston Wilson*
Treasurers: *Walter McNeill, Jimmy Moffitt*
Auditor: *Dean Watson*
Exec. Committee:

D. Blair, B. Burns, E. Crampton, W. Leeburn, J. Loughrey, J. Magee, G. McConkey, H. McConnell, J. McWhirter, S. Patterson, M. Scott, R. Shaw, L. Swinerton, C. Harpur, T. Johnston, M. Shillington, J. Sheeran, R. Mcpherson, J. Holland, D. Saunders, J. Ferguson, L. Simms

OBJECTS OF THE ASSOCIATION

The objects of the Association shall be to unite beekeepers for their mutual benefit to serve the best interests of beekeeping by all means within its power and to foster its healthy development.

For the purpose of achieving these objects the Association will:

- promote the formation of local beekeepers' Associations
- disseminate information and advice about beekeeping
- provide examination facilities in the art of beekeeping
- maintain and improve the bee environment.

SECRETARIES OF ASSOCIATIONS

Ballymena	*J. McWhirter*, 67 Main Street, Cullybackey, Co. Antrim
Belfast	*Sam Patterson*, 17 Donaghadee Road, Mioisle, Co. Down
Derg & Mourne	*Mrs. Joan Ferguson*, 21 Strabane Road, Newtownstewart, Co. Tyrone BT78 4AZ
Dromore	*R. Shaw*, 50 Pine Cross, Seymour Hill, Dunmurry
East Antrim	*Dr. D.H. Saville C. Biol., M.I. Biol.*, 20 Hollow Road, Islandmagee BT40 3RL
Enniskillen	*H. McConnell*, Old School House, Brookeborough, Enniskillen, Co. Fermanagh
Londonderry	*Liam Salmon*, 23 Kingsford Park, Culmore Road, Co. Londonderry
Randalstown	*Walter McNeill*, 10 Masserene Gardens, Antrim ☎ (08494) 64648
Roe Valley	*Jim Loughrey*, 169 Roemill Road, Limavady, Co. Londonderry ☎ (05047) 64864
Warrenpoint	*Mrs. Anne Smyth*, 30 The Hill, Srinan Road, Newry BT34 2PJ

WELSH BEEKEEPERS' ASSOCIATION
CYMDEITHAS GWENYNWYR CYMRU

General Secretary:
> *B.D. Rowlands* BSc.
> Trem y Clawdd
> Fron Isaf, Y Waun
> Clwyd   LL14 5AH
> ☎ (01691) 773300

President: *Cdr. E.G. Verge*, Derwen Fach, Llandygwydd, Aberteifi   SA43 2QU
Vice Presidents: *C. Davies*, 40 Heol Alltiago, Pontarddulais, Abertawe   SA4 1HU
> *W.J. Jones*, Avalon, Llanfrothen   Ll48 6LJ
Chairman: *J.A. Hall*, Cnwc Sych, Llanfair Clydogau, Llanbedr Pont Steffan   SA48 8LJ
> ☎ (01570) 493350
Vice Chairman: *C. Wynne Jones*, Ty Brith, Pentre Celyn, Rhuthun   LL15 2SR
> ☎ (01978) 790279
Treasurer: *Mrs. E. Edwards*, The Post Office, 74 Vincent Street, St. Helens, Merseyside
> WA10 1LD   ☎ (01744) 22098
Editor: *Miss A.L. Jones* B.Sc., 13 Maes Meugan, Llanrhydd, Rhuthun   LL15 1YH
> ☎ (01824) 703523
Labels: *Marged Phillips*, Pencefn, Tynreithin, Tregaron   SY25 6LL
Insurance: *F. Brown*, Vale Mount, Gellifor, Rhuthun, Clywd   LL15 1RY
> ☎ (01824) 70427
Legal Adviser: *K. Williams* LLB, 1 Sgwar Harford, Llanbedr Pont Steffan   SA48 7HD
Promotions: *Mr. & Mrs. R. Mendy*, Hill House Farm, Pencraig, Powys   LD8 2UN
> ☎ (01544) 230175
Audio-Visual Aids Secretary:
> *F. Eckton*, Cartref, Llanafan Fawr, Llanfair Ym Muallt   LD2 3LT
> ☎ (01554) 753451

EMERGENCY COMMITTEE
General Secretary, Treasurer, President, Chairperson, Vice chairperson

EXAMINATIONS BOARD
Secretary: *D.H. Ferguson Thomas*, Erw Lon, Llanwrda   SA19 8HD   ☎ (01550) 777132
> *C. Davies*
> *D.H. Jones*
> *W.J. Jones*
> *P.A. Gregory (Co-opted)*

EDITORIAL BOARD
Chairman, Vice Chairman, Secretary, Treasurer, Editor, Assistant Editors

> R. Mendy,
> P. Mendy (Co-opted)
> R. Jones
> D.H. Ferguson Thomas

EDUCATION, RESEARCH & DEVELOPMENT COMMITTEE
Chairman, Vice Chairman, Secretary, Treasurer

> P. Gregory (Co-opted)
> C. Davies
> L. Chirnside

WELSH NATIONAL HONEY SHOW COMMITTEE

> J. Thomas CBE
> C. & J. Davies
> G. & M. Morgan
> D.H. & M.M. Jones
> W.J. Jones
> B.D. Rowlands
> F. Eckton
> T. Davies
> J.E. Williams
> Mr. and Mrs. I. Richards
> M. and D. Charles
> D.H. Ferguson Thomas
> Mrs. P. Parry
> E. Hughes

CONVENTION COMMITTEE
Secretary: Mr. H. Taylor, Green Acres, Clayford Road, Cilgeti SA68 0RR
☎ (01834) 813775

Trade Stand Secretary:
> R. Mendy, Hill House Farm, Pencraig, Powys LD8 2UN
> ☎ (01544) 230175
> J. Burgess
> L. Chirnside
> I. Evans
> R. Mendy

PAST PRESIDENTS

> G. Gilbert
> H.A. Peter
> D.H. Jones

HONORARY LIFE MEMBERS
D.H. Jones   Capel Cynon, Llandysul, Dyfed
G.D. Gilbert   Illtyd's Well, Llanrhidian, Abertawe
Dr. J.J. Marsden   Aberhonddu, Powys

R.H. *Bowering* Sunny Bank, Aberhonddu, Powys
H.A. *Peter* MA ᶜ/ₒ Crossfield House, Rhaeadr, Powys
Bro. *Adam* OSB OBE
St. Mary's Abbey, Buckfastleigh, Dyfnaint

AIMS OF THE ASSOCIATION
• To promote and develop beekeeping in Wales.
• To conduct examinations in beekeeping.
• To liaise with organisations, institutions and bodies for the benefit of beekeeping in Wales.

EXAMINATIONS
The Board conducts six grades of examinations: Junior, Primary, Intermediate, Practical, Honey Show Judges, Senior. Candidates following the Duke of Edinburgh Award Scheme may apply to the Examinations Board Secretary for information regarding the inclusion of beekeeping in their course submissions.

WELSH NATIONAL HONEY SHOW
This is a four day event, held annually during July at the Royal Welsh Agricultural Society Showground, Llanelwedd, Powys. grid ref: S0 040520. Schedules may be obtained from the Secretary, WNHS Committee, RWAS, Llanelwedd   LD2 3SY.

Affiliated associations support the honey sections of their local honey shows. Information is available from local association secretaries.

INSURANCE
All individual and fully paid up members of beekeeping associations affiliated to WBKA are covered against 'Public and Product' liability claims. All affiliated associations are covered against public liability during conventions officially organised by the association.

Paid up members may further insure their equipment and stocks to cover loss and damage caused by fire, storm, flood, theft, accidental or malicious damage. Details are obtainable from the Insurance Secretary.

Cover against losses due to Foul Brood diseases is provided for individual members. Affiliated Associations provide this cover for its members.

LIBRARY
The reference sections of all county libraries in Wales have details of the names and addresses of Secretaries of Associations affiliated to WBKA. Books on beekeeping can be borrowed from county, branch and mobile libraries. The Library, Ffordd y Bala, Dolgellau LL40 2YS, has been nominated to stock beekeeping books. Members of associations affiliated to IBRA may borrow books/documents from its library.

• *NEWYDDLEN* - Quarterly Bulletin
  This bulletin is published during March, June, September and December and is provided free to individual members and members of affiliated associations.

WBKA ANNUAL CONVENTION
This event is held at the Royal Welsh Agricultural Society Showground, Llanelwedd, Llanfair ym Muallt, Powys, grid ref: SO 040520 during late March or early April.

SALES

Publications, Honey Show prize cards, ties, badges etc. are available at the Welsh National Honey show and the Convention or from the promotions secretary. Honey jar labels (in Welsh and English) are available from the Labels Secretary.

AFFILIATED ASSOCIATIONS AND SECRETARIES

Aberconwy   *P. McFadden*, Llwyn, Rowen, Conwy   LL32 8YP   ☎ (01492) 650851
Aberystwyth & District/Aberystwyth a'r Cylch
          P.D. Mathys, 1 Station Cottage, Ystrad Meurig, Pont rhyd Fendigaid
          SY25 6AX   ☎ (01974) 831451
Anglesey/Ynys Mon
          M.W. Shaw, Llwyn Ysgaw, Dwyran, Llanfairpwll, Gwynedd   LL61 6RH
          ☎ (01248) 430811
Brecknock & Radnor/Brycheiniog a Faesyfed
          Mrs. V. Veasey, Pen Cerrig Home Farm, Llanelwedd, Llanfair ym Muallt
          LD2 3LT   ☎ (01982) 553177
Cardiff & District/Caerdydd a'r Cylch
          W.B. Lee, Croes Heol, Llanmaes, De Morgannwg   CF6 2RT
          ☎ (01446) 750072
Carmarthen/Caerfyrddin
          Mrs. M. Macleod, Pant Melyn, Pen y Bont, Caerfyrddin   SA33 6QN
          ☎ (01994) 8473
East Camarthen/Dwyrain Caerfyrddin
          A. Surman, Trem y Berllan, Tal y Llechau, Llandeilo   SA19 7YH
          ☎ (01558) 685422
Gogledd Dyfed *W.I. Griffiths*, Llaindeg, Commins Coch, Aberystwyth   SY23 3BG
          ☎ (01970) 623334
Lampeter & District/Llanbedr a'r Cylch
          Mrs. E. Evans, Glandular House, LLanybydder   SA40 9RN
          ☎ (01570) 480531
Llyn ac Eifion   *J.G. Roberts*, 22 Stryd Lombard, Porthmadog   LL49 9AP
          ☎ (01766) 512847
Mid Glamorgan/Morgannwg Ganol
          Mrs. S.E. Verran, 4 The Moulders, Abercynffyg, Pen y bont ar Ogwr,
          Morgannwg Ganol   CF32 9AH   ☎ (01656) 724249
Montgomery/Trefaldwyn
          P. Thomas, 23 Bryn Meadows, Y Drenewydd   SY16 2DS   ☎ (01686) 627315
North Clwyd/Gogledd Clwyd
          Mrs. P. Parry, 9 Heol Wen, Coedpoeth, Wrecsam   LL11 3HD
          ☎ (01978) 753908
Pembrokeshire/Sir Benfro
          Mr. H. Jenkins, Steynton Cottage, Merlins Cross, Penfro   SA71 4AG
          ☎ (01646) 684053
South Clwyd/De Clwyd
          M. Simister, The Cottage, Marford Mill, Yr Orsedd, Wrecsam   LL12 0HL
          ☎ (01244) 570106

Swansea & District/Abertawe a'r Cylch
H.T. *Edmunds*, 1A Aylesbury Road, Brynmill, Abertawe   SA2 OBS
☎ (01792) 649968

Teifiside/Glannau Teifi
Mrs. *F. Lee*, Glas y Dorlan, Maes y Meillion, Llandysul   SA44 4NQ
☎ (0154) 555281

West Glamorgan/Gorllewin Morgannwg
Mrs. *M. Blackmore*, 7 Heol Hendrefoilan, Ty Coch, Abertawe   SA2 9LS
☎ (01792) 298042

NON AFFILIATED
Abergavenny/Y Fenni
R. *Sadler*, Harvest Hill, The Narth, Trefynwy   NP5 4QH
☎ (01600) 860527

Gwent   P. *Hayward*, Llana Nant Farm, Penallt, Trefynwy   NP5 4AP
☎ (01600) 712864

Meirionydd   Michael *Furness*, Tyddyn Cwper, Gellilydan   LL41 4EP   ☎ (01766) 85353

Newport/Casnewydd
T. *Heaton*, 2 Coed Garw, Croes y Ceiliog, Cwmbran   NP44 2NJ
☎ (01633) 867601

WBKA QUALIFIED HONEY SHOW JUDGES

D.H. Jones	Bryn Cynon, Capel Cynon, Llandysul SA44 4TJ
W.J. Jones	Avalon, Llanfrothen, Gwynedd LL48 6LJ
G.J. Hartshorn	Maesteg, Y Stryd Fawr, Llanberis LL55 4HB
T. Collins NDB	Neuaddlwyd, Ciliau Aeron, Llanbedr pont Steffan SA48 7RE
C. Davies	40 Heol Alltiago, Pontarddulais, Abertawe SA4 1HU
H. Taylor	Greenacres, Clayford Road, Cilgeti SA68 0RR
M. Bessent	Gwili Lodge, Heol Lot Wen, Cwm Gwili, Rhydaman SA18 3RP
L. Chirnside	Bryn Y Pant Cottage, Upper Llanofer, Y Fenni NP7 9EW
D.H. Ferguson Thomas	
	Erw Lon, Llanwrda SA19 8HD
Mrs. D. Sweet	5 Parima Road, Valsayn South, Trinidad, West Indies
	or (ᶜ/₀ 19 Millwood, Llys faen, Caerdydd)

NON-PRACTISING

E.B. Plumb	7 Ovington Terrace, Caerdydd CF5 1GF
R. Williams	24 Bron y Clwyd, Llanfair D C, Rhuthun, Clwyd LL15 2SB

CENTRAL SCIENCE LABORATORY NATIONAL BEE UNIT

Head of Unit: *Medwin Bew*
Central Science Laboratory
MAFF Luddington
Stratford-upon-Avon
CV37 9SJ
☎ (01789) 750601 **Fax:** (01789) 750957

PRESENT PERMANENT TECHNICAL STAFF

Beekeeping Inspectors Technical Manager:	*Mike Brown*
Laboratory Technical Manager:	*Melanie Hughes*
Varroa Research:	*Stephen Martin*
I.T. Officer:	*David Wilkinson*
Scientific Officers:	*Sarah Smith*
	Julian Perrett
Assistant Scientific Officers:	*Amelia Jones*
	Paul Wilkins
	Nicola Hatton

The CSL National Bee Unit is part of the MAFF Central Science Laboratory research agency. It provides diagnostic, consultancy and research services to MAFF, commerce and to beekeepers, both amatuer and professional. The Unit has modern facilities, including laboratories with first class computer support, as well as 150 bee colonies and the apiary buildings to support them. The Unit laboratories work to the OECD Principles of Good Laboratory Practice to ensure a high professional standard and, as well as being scientists, all technical staff are trained, practical beekeepers. The Unit is also supported by analytical chemists and agricultural and envirionmental specialists in the rest of CSL.

STATUTORY WORK

The Unit examines thousands of samples of dead bees, brood combs and hive debris each year for evidence of serious pests and diseases, e.g. American & European foul brood (bacterial diseases of brood) and varroa mites. Much of this work, conducted under EC directive EC/65/92 and our own Bees Act 1980, is funded by MAFF, without charge to beekeepers.

The Unit is also the authorised MAFF facility for examination of queen bee consignments imported from countries outside the EC. (*Prospective importers should contact MAFF Horticulture & Potatoes Division, Ergon House, 17 Smith Square, London SW1P 3JR, for further information*)

Since April 1994, the Unit has had responsibility for the MAFF bee health inspection programme and employ 10 full-time Regional Bee Inspectors and a further 22 Seasonal Bee Inspectors throughout England and Wales for AFB, EFB and varroa mite inspections of beekeepers' colonies. They are also employed as bee health advisers to beekeepers, providing leaflets, talks and workshops, for example, on the control and treatment of serious diseases of honey bees. This work is funded by MAFF Horticulture & Potatoes Division and carried out under the Bees Act 1980.

REGIONAL BEE INSPECTORS

Northern - Cumbria, Lancashire, Northumberland, Tyne & Wear
> Mr. I. McLean, Asland, Flash Lane, Rufford, Ormskirk, Lancashire
> L40 1SW ☎ (01704) 822831

Western - Merseyside, Greater Manchester, Cheshire, Staffordshire, Shropshire, West Midlands, Warwickshire
> Mr. D. Wright, 5 Briar Close, Walton-on-the-Hill, Stafford, Staffordshire
> ST17 0NG ☎ (01785) 661891

Severn - Hereford & Worcestershire, Avon, Somerset, Gloucester
> Mr. L.J. Dixon, The Square, Titley, Nr. Kington, Herefordshire   HR5 3RG
> ☎ (01544) 231245

South West - Devon, Cornwall
> Mr. L. Davie, Nymet Cottage, Walson, Bow, Crediton, Devon   EX17 6JX
> ☎ (01363) 82399

Southern - Oxfordshire, Berkshire, Buckinghamshire, Wiltshire, Dorset, Hampshire
> Dr. B. Cullen, 26 Sweetcroft Lane, Hillingdon, Middlesex   UB10 9LD
> ☎ (01895) 810649

South Eastern - Greater London, Kent, Surrey, West Sussex, East Sussex
> Mr. J. Morton, 99 Burket Close, Norwood Green, Middlesex   UB2 5NT
> ☎ (0181) 893 8735

Eastern - Bedfordshire, Hertfordshire, Cambridgeshire, Essex, Suffolk, Norfolk
> Mr. J. Blakesly, Bush Farm, Somersham Road, Flowton, Ipswich, Suffolk
> IP8 4LN ☎ (01473) 658078

East Midlands - Northamptonshire, Leicestershire, Derbyshire, Nottinghamshire, Lincolnshire
> Mr. D. Kemp, 332 Spring Lane, Mapperley Plains, Mapperley, Nottingham
> NG3 5RQ ☎ (01159) 661238

North East - South Yorkshire, West Yorkshire, North Yorkshire, Humberside, Durham, Cleveland
> Mr. J.M. Goodman, East Nettlepot, Lunedale, Middleton-in-Teesdale,
> Co. Durham   DL12 0NX ☎ (01833) 640879

Wales - Clwyd, Gwynedd, Powys, Mid. & South Glamorgan, Gwent, Dyfed
> Mrs. P.M. Gregory, Pentrebwlen, Llanddewi Brefi, Tregaron, Dyfed
> SY25 6PA ☎ (01570) 493601

PESTICIDE INCIDENTS

The Unit participates, with other CSL and MAFF specialists, in the Wildlife Incident Investigation Scheme, which monitors the environmental effect of pesticides on wildlife, including bees, on behalf of MAFF. Samples of bees from suspected pesticide incidents, sent in to the Unit by beekeepers, are analysed for cause of death. The results are used to prevent recurrence of problems wherever possible. This work is presently conducted without charge to the beekeeper.

PESTICIDE TOXICITY

The Unit has a well-developed expertise in testing pesticide toxicity to honeybees to formal standards, e.g. European Plant Protection Organisation (EPPO) and the UK Control of Pesticides Regulations 1986. These include laboratory and field tests to assess the toxicity of compounds to both adult bees and broods.

RESEARCH AND DEVELOPMENT

Currently the Unit's research concentrates on the biology of the varroa mite infestations in British conditions so that more effective and appropriate diagnostic and monitoring tools may be made available to beekeepers. This work is funded by MAFF with a contribution from the British Beekeepers' Association.

Commercial research is also undertaken, e.g. testing new beekeeping appliances and feedstuffs and establishing the geographical and botanical origins of pollen in honey.

CONSULTANCY

Technical advice is provided in all aspects of apiary managment and bee husbandry to beekeeprs and commercial and other organisations with an interest in beekeeping, e.g. bee farm planning appraisals.

The Unit has long established links with many European and North American research centres, beekeeping specialists and beekeeprs to ensure an up to date knowledge of current beekeeping trends and research. It is also involved with training overseas bee specialists.

MOVE TO YORK

In autumn 1996 the Unit will be relocating with the Central Science Laboratory to York.

DEPARTMENT OF AGRICULTURE FOR NORTHERN IRELAND

Beekeeping Officer for the Department:

William P. Duff
Agriculture Technology Division
Greenmount College of Agriculture and Horticulture
22 Greenmount Rd.
Antrim
BT41 4PU
☎ (01849) 462114

This was another difficult year for beekeepers in Northern Ireland. Honey Production was low and loss of colonies high due to adverse weather conditions.
Steps are being taken in conjunction with UBKA to reverse this trend.
On the positive side, the autumn survey of hives across the province revealed no cases of Varroasis.

During the Autumn/Winter of 1994/95 the Department organised:

• Two "Introductory courses in beekeeping" each consisting of 10 evening classes. 55 people attended.

SCOTTISH AGRICULTURAL SCIENCE AGENCY

THE SCOTTISH OFFICE
AGRICULTURE AND FISHERIES DEPARTMENT

Headquarters: Pentland House
47 Robbs Loan
Edinburgh
EH14 1TY

Executive Agency: Scottish Agricultural Science Agency
East Craigs
Edinburgh
EH12 8NJ

Bee Diseases: Plant Health Section
☎ (0131) 244 8863

Pesticide Incidents: *Elizabeth Sharp*
Chemistry Section
☎ (0131) 244 8874

The Scottish Office Agriculture and Fisheries Department (SOAFD) has responsibility for policy matters, the control of the movement of bees in the event of an outbreak of a statutory notifiable disease, maintaining records of disease incidents, issuing import licences for queen bees, initiating enforcement action where bee mortalities have resulted from the misuse or abuse of pesticides and for liaison with the Scottish Beekeepers Association.

SOAFD Agricultural Officers, located at area offices throughout Scotland carry out field inspections following notification of bee mortalities resulting from the misuse of pesticides. Those Officers who are Bees Officers investigate notified incidents of statutory bee diseases.

The Scottish Agricultural Science Agency (SASA) has responsibility for the analysis of pesticides in dead bees under the Control of Pesticides Regulations and for the diagnosis of statutory bee diseases under the Bee Diseases Control Order 1982 and the Importation of Bees Order 1980.

BEE DISEASES
Under the Bee Diseases Control Order 1982, beekeepers are required to notify SOAFD of suspected cases of foul brood diseases or varroasis. These should be notifed to the Principal Agricultural Officer (PAO) at the nearest SOAFD Area Office who will arrange for a Bees Officer to carry out an inspection. Brood combs or hive debris from colonies thought to be infected will be sent to SASA for laboratory examination. Alternatively, a beekeeper may send

hive debris direct to Plant Health Section, (Bee Diseases), SASA, at the above address for examination for the parasitic mite, *Varroa jacobsoni.*

Following the detection of varroa in Devon in April 1992, widespread searches of bee colonies were carried out in spring and autumn in Great Britain to assess the spread of the disease. No mites were found in either of these searches in Scotland.

The area of England affected by the mite has been declared a Statutory Infected Area (SIA), and the movement of colonies and queens has been under restriction. To prevent the spread of infection into Scotland, only queen bees with attendant workers may be moved exceptionally under licence.

Queen bees and their attendant workers entering Scotland under licence issued under both the Importation of Bees Order 1980 (from outwith Great Britain) and the Bee Diseases Control Order (from the SIA) are examined for varroa mite infestation. Queen bees are inspected by the Bees Officer at the premises of the importer. Attendant workers are sent to SASA East Craigs for examination.

No charges are made for these services.

PESTICIDE INCIDENTS

As part of the Wildlife Incident Investigation Scheme, SASA undertakes analytical investigations into bee mortalities where pesticide poisoning may have been involved. Beekeepers should send samples of dead bees (200) direct to SASA, Chemistry Section, for analysis. In the case of major incidents, beekeepers are advised to contact the PAO at the nearest Area Office so that an early field investigation can be instigated. A report will be issued to the beekeeper and if appropriate, enforcement action under the Control of Pesticides Regulations 1986 will be initiated.

No charges are made for these services.

REPORT FOR 1994

Bee Disease Control Order 1982

American Foul Brood: - 2 incidents

European Foul Brood: - No incidents

Licensed movement of queen bees from SIA
                         - 5

Examination of hive inserts for *Varroa jacobsoni* following diagnosis using Bayvarol:
                    - SOAFD Spring search - 130 colonies from 22 beekeepers
                    - SOAFD Autumn search - 464 colonies from 112 beekeepers

Examination of hive debris and inserts for *Varroa jacobsoni* following various methods of detection: (sent direct to SASA by beekeepers)
- 121 samples from 87 beekeepers

No Varroa jacobsoni mites were found in any of the samples examined.

Importation of Bees Order 1980
- none

Pesticide Incidents	- Number reported	8
	- number positive	2
	- Pesticides involved	gamma HCH and dieldrin
		bendiocarb

THE SCOTTISH OFFICE AGRICULTURE & FISHERIES DEPT. AREA OFFICES

All communications should be addressed to:
The Principal Agricultural Officer.

Angus/N.E. Fife - Dundee Area Office:
Northern College of Ed Buildings, Gardyne Road, Broughty Ferry, Dundee DD5 1PE ☎ (01382) 462840 **Fax:** (01382) 454128

Argyll & Western Isles - Oban Area Office:
Cameron House, Albany Street, Oban PA34 4AE
☎ (01631) 563071 **Fax:** (01631) 566756

Borders - Galashiels Area Office:
Cotgreen Road, Tweedbank, Galashiels TD1 3SG
☎ (01896) 758333 **Fax:** (01896) 756803

Grampian - Inverurie Area Office:
Thainstone Court, By Inverurie, Aberdeenshire AB51 5YA
☎ (01467) 626222 **Fax:** (01467) 626217

Highland - Inverness Area Office:
Longman House, 28 Longman Road, Inverness IV1 1SF
☎ (01463) 234141 **Fax:** (01463) 714697

Northern - Thurso Area Office:
Strathbeg House, Clarence Street, Thurso KW14 7JS
☎ (01847) 893104 **Fax:** (01847) 895983

Northern Isles - Kirkwall Area Office:
Tankerness Lane, Kirkwall, Orkney KW15 1AQ
☎ (01856) 875444 **Fax:** (01856) 873309

Perth and Kinross - Perth Area Office:
1 Mill Street, Perth PH1 5HZ
☎ (01738) 443266 **Fax:** (01738) 630751

Southern - Dumfries Area Office:
161 Brooms Road, Dumfries DG1 3ES
☎ (01387) 255292 **Fax:** (01387) 250497

South Western - Ayr Area Office:
Russell House, King Street, Ayr KA8 0BE
☎ (01292) 610188 **Fax:** (01292) 611483

Hamilton - Sub Office:
Cadzow Court, 3 Wellhall Road, Hamilton ML3 9BG
☎ (01698) 281166 **Fax:** (01698) 285277

Sterling - Sub Office:
Government Buildings, 2 St. Ninians Road, Stirling FK8 2HR
☎ (01786) 473272 **Fax:** (01786) 465033

BEE DIAGNOSTIC SERVICE FOR THE REPUBLIC OF IRELAND (*TEAGASC*)

Agricultural and Food Development Authority
Clonroche Research Centre
Co. Wexford
Ireland

Patsy Bennett: ☎ (054) 44106

ANNUAL REPORT 1992

The Bee Diagnostic Service was transferred from Kildalton College to Clonroche on 1st March 1992. A total of 195 beekeepers sent in 368 samples. Analysis of the samples received showed that Acarine and Nosema are still a common problem with bee colonies. The postal strike had a serious effect on the number of samples received as it coincided with the main examination period in beekeeping.

VARROASIS

The discovery of the mite *Varroa jacobsoni* in Devon in Southern England in early April of 1992 brings the threat of varroasis ever closer. Once the mite becomes established it cannot be eradicated and will eventually spread to all apiaries. However, it can be controlled by biotechnical or chemical treatments to enable beekeeping to be carried on. These methods involve extra manipulations and the chemical treatments are expensive.

The Department of Agriculture, Fisheries and Food have imposed a total ban on the importation of bees and queens to ensure that the mite is not brought into the country. In the meantime beekeepers should monitor their bee colonies by using hive inserts and the tobacco treatment to ascertain that the mite is not already here. A sample of the resulting debris should be sent to Richard Dunne, Kinsealy Research Centre, Malahide Road, Dublin 17, for analysis.

FORWARDING OF SAMPLES

Samples of live or recently dead bees should be sent to Clonroche in **match boxes** or similar type **cardboard boxes**. In spite of repeated requests not to send samples in plastic or glass containers we still receive samples in both and this makes the examination of these extremely difficult. Send 25 to 30 bees in each sample.

Comb samples should be wrapped in newspaper and enclosed in fairly rigid containers. Do not send comb samples in plastic bags as this leads to rapid fungal growth in warm weather. Label all samples clearly and do not forget to include your name and address. All samples (except hive debris) should be accompanied by the appropriate fee. In cases where AFB or EFB are diagnosed the fee will be refunded.

ANNUAL REPORT 1992	1992	1991	1990
No. of Beekeepers	195	293	271
No. of Bees examined	10696	13443	12620
No. of Colonies diagnosed	368	525	491
Adult Diseases:			
Nosema	46	85	33
Amoeba	2	8	3
Acarine	40	48	36
Brood Diseases:			
No. of Brood Combs received	45	90	86
America Foul Brood	3	17	27
European Foul Brood	0	0	0
Sac Brood	0	1	0
Chalk Brood	14	18	15
Chilled Brood	0	14	17
Addled Brood	0	0	0
Mouldy Pollen	5	10	3
Starvation	2	2	2
Wax Moth	1	3	2
Braula Coeca	0	3	2
Toxicity	2	5	4
Pollen Mites	0	0	0
Non specific Dysentry	0	0	0
Neglected Brood	4	0	0

BEEKEEPING AND BEE HEALTH STATISTICS 1993

MAFF Horticulture & Potatoes Branch A

TABLE I

Number of recorded beekeepers and colonies with analysis of beekeepers by numbers of colonies owned.

	No. of Beekeepers with live colonies					Total no. of Colonies	No. of Colonies owned by beekeepers with 40 or more colonies
	Owning 1 to 10 colonies	Owning 11 to 39 colonies	Owning 40 to 100 colonies	Owning over 100 colonies	Total		
ENGLAND							
1991	26,216	2,350	290	91	28,947	152,216	31,796
1992	19,184	1,890	222	87	21,383	138,354	27,434
1993	25,273	2,239	244	84	27,840	142,663	24,552
WALES							
1991	2,097	203	29	6	2,335	14,093	1,831
1992	2,141	211	23	9	2,384	13,570	2,547
1993	2,142	211	23	9	2,385	13,621	2,547
ENGLAND AND WALES							
1991	28,313	2,553	319	97	31,282	166,309	33,627
1992	21,325	2,101	245	96	23,767	151,924	29,981
1993	27,415	2,450	267	93	30,225	156,284	27,099

NB - There is no official registration of beekeepers and as not all apiaries are inspected each year these figures are necessarily only estimates.

TABLE 2

Inspections under the Bee Disease Order 1982.

	Number of colonies inspected			Number of diseased colonies destroyed or treated in the year			Number of diseased colonies destroyed or treated as percentage of colonies inspected	
	LIVING	DEAD	TOTAL	AFB	EFB	TOTAL	AFB	EFB
ENGLAND								
1991	29,960	2,553	32,513	225	795	1,028	0.69	2.17
1992	26,593	2,207	28,800	117	578	695	0.41	2.01
1993	28,667	1,937	30,604	98	539	637	0.32	1.76
WALES								
1991	2,639	139	2,778	41	14	55	1.48	0.47
1992	2,843	218	3,061	10	23	33	0.33	0.75
1993	3,084	309	3,393	12	17	29	0.35	0.50
ENGLAND AND WALES								
1991	32,599	2,692	35,291	266	809	1,075	0.75	2.31
1992	29,436	2,425	31,861	127	601	728	0.40	2.28
1993	31,751	2,246	33,997	110	556	666	0.32	1.64

TABLE 3

Results of laboratory examination of sample combs.
Service provided by the Central Science Laboratory National Bee Unit.

LABORATORY SERVICE PROVIDED BY Central Science Laboratory National Bee Unit	NUMBER OF COMBS RECEIVED	LABORATORY DIAGNOSIS		
		AFB	EFB	NO DISEASE
Samples submitted by Bee Officers				
1991	1,040	80	791	169
1992	765	57	586	122
1993	797	76	553	168
Samples submitted by Beekeepers				
1991	67	8	18	41
1992	67	3	15	49
1993	34	2	3	29
Total England and Wales				
1991	1,107	88	809	210
1992	832	60	601	171
1993	831	78	556	197

TABLE 4 - Breakdown of Inspections and Analysis by County 1993

COUNTY	No. of Bee-keepers	No. of Live Colonies	No. of Colonies Examined	Samples sent to Laboratory		AFB No. of Colonies Distroyed		EFB		Total No. of Diseased Colonies	No. of Diseased Apiaries
				Bees Officers	Bee Keepers	Field Diagnosis	Lab. Diagnosis	No. of Colonies Distroyed	No. of Colonies Treated		
ENGLAND											
Avon	587	2120	485	6	5	0	2	0	1	3	3
Bedfordshire	324	2158	771	0	0	0	0	0	0	0	0
Berkshire	335	1432	334	63	2	0	0	15	20	35	19
Buckinghamshire	429	1810	265	6	1	0	2	1	1	4	4
Cambridgeshire	637	3857	2010	7	0	1	1	1	1	4	4
Cheshire	357	2337	313	9	0	0	0	0	6	6	2
Cleveland	76	415	225	0	0	0	0	0	0	0	0
Cornwall	2265	8343	840	27	2	10	1	14	11	36	23
Cumbria	472	1661	140	0	0	0	0	0	0	0	0
Derbyshire	309	1620	485	1	0	0	0	0	1	1	1
Devon	1872	11307	1047	56	2	16	3	25	20	64	19
Dorset	891	5245	988	64	0	0	4	4	54	62	62
Durham	179	955	326	0	0	0	0	0	0	0	0
E. Sussex	512	2070	365	1	0	0	0	0	0	0	0
Essex	704	4580	1525	26	1	0	11	3	4	18	14
Gloucestershire	592	3831	1068	16	0	0	0	3	13	16	10
Greater London	1085	3244	818	120	3	3	10	39	48	100	45
Greater Manchester MC	344	255	259	4	0	0	2	0	0	2	1
Hampshire	1046	4185	764	37	0	0	0	17	19	36	20
Hereford & Worcester	850	6492	2190	57	1	0	5	7	43	55	25
Hertfordshire	431	2673	866	24	1	0	0	5	14	19	14
Humberside	389	2360	278	0	0	0	0	0	0	0	0
Isle of Wight	138	487	27	0	1	0	0	0	0	0	0
Kent	1059	5594	1505	93	0	4	5	31	51	91	51
Lancashire	333	1016	279	1	1	2	1	0	0	3	3
Leicestershire	379	2205	161	3	0	0	0	0	2	2	1
Lincolnshire	655	4956	1428	1	0	0	0	1	0	1	1
Merseyside MC	235	311	314	7	0	0	6	0	0	6	3
Norfolk	1070	5686	1339	29	0	0	0	7	10	17	17
N. Yorkshire	819	4199	549	0	2	0	0	0	0	0	0
Northamptonshire	356	2171	524	1	0	0	1	0	0	1	1
Northumberland	358	2771	592	1	0	0	1	0	0	1	1
Nottinghamshire	322	2194	1002	0	0	0	0	0	0	0	0
Oxfordshire	330	2119	538	24	7	0	6	6	14	26	13
Shropshire	720	4085	689	4	0	0	1	0	2	3	3
Somerset	1033	4148	738	14	0	0	2	4	3	9	9
S. Yorkshire MC	210	981	466	1	0	0	0	0	0	0	0
Staffordshire	527	2691	319	4	0	0	0	2	0	2	2
Suffolk	862	4673	1827	5	0	2	1	0	0	3	3
Surrey	995	3007	294	24	1	0	3	2	6	11	8
Tyne & Wear MC	126	488	135	0	0	0	0	0	0	0	0
Warwickshire	355	4698	367	11	2	0	0	1	9	10	4
West Midlands MC	447	2218	446	1	1	1	1	0	0	2	1
West Sussex	520	3293	453	10	1	5	3	0	0	8	4
W. Yorkshire	374	1914	0	0	0	0	0	0	0	0	0
Wiltshire	786	3808	604	9	0	0	0	2	3	5	5
Total for England	**27,695**	**142,663**	**30,958**	**767**	**34**	**44**	**72**	**190**	**356**	**662**	**396**
WALES											
Clwyd	347	1561	676	0	0	3	0	0	0	3	0
Dyfed	911	5728	806	9	0	5	5	0	1	11	5
Gwent	106	997	208	3	0	0	0	0	0	0	0
Gwynedd	372	1559	790	2	0	0	1	0	0	1	1
Powys	436	1494	397	9	0	0	0	6	3	9	2
Mid Glamorgan	74	618	248	1	0	0	0	0	0	0	0
S. Glamorgan	97	561	58	1	0	0	0	0	0	0	0
W. Glamorgan	94	1103	110	5	0	0	0	0	0	0	0
Total for Wales	**2,437**	**13,621**	**3,293**	**30**	**0**	**8**	**6**	**6**	**4**	**24**	**11**
Total for England & Wales	**30,132**	**156,284**	**34,251**	**797**	**34**	**52**	**78**	**196**	**360**	**686**	**407**

TABLE 5 - Voluntary Varroa Search England & Wales 1993

ENGLAND	No. of Beekeepers	No. of Colonies Examined	No. of Apiaries Sampled	Varroa Positive Apiaries
Avon	91	234	119	18
Bedfordshire	17	58	20	0
Berkshire	61	172	77	5
Buckinghamshire	62	246	81	3
Cambridgeshire	12	64	25	0
Cheshire	24	71	32	0
Cleveland	20	41	24	0
Cornwall	53	158	71	6
Cumbria	17	61	26	0
Derbyshire	44	145	56	0
Devon	118	424	180	93
Dorset	123	399	188	69
Durham	16	155	26	0
E. Sussex	41	151	52	3
Essex	76	252	110	0
Gloucestershire	26	123	37	0
Greater London	71	162	81	21
Greater Manchester	6	12	8	0
Hampshire	232	769	315	47
Hereford & Worcester	52	165	76	0
Hertfordshire	38	146	57	0
Humberside	18	41	21	0
Isle of Wight	37	260	77	34
Kent	128	416	168	38
Lancashire	13	25	13	0
Leicestershire	19	56	24	0
Lincolnshire	69	225	97	0
Merseyside	12	41	13	0
Norfolk	53	213	72	0
N. Yorkshire	42	176	56	0
Northamptonshire	16	55	17	0
Northumberland	16	80	21	0
Nottinghamshire	24	87	28	0
Oxfordshire	38	122	44	1
Shropshire	17	62	18	0
Somerset	131	646	194	34
S. Yorkshire	8	18	10	0
Staffordshire	43	129	57	0
Suffolk	36	114	45	6
Surrey	118	324	174	51
Tyne & Wear	12	22	12	0
Warwickshire	27	371	59	0
West Midlands	53	168	75	0
West Sussex	110	361	150	7
W. Yorkshire	19	74	21	0
Wiltshire	103	303	147	5
Totals	**2,362**	**8,397**	**3,274**	**441**

WALES	No. of Beekeepers	No. of Colonies Examined	No. of Apiaries Sampled	Varroa Positive Apiaries
Clwyd	15	29	15	0
Dyfed	18	79	20	0
Gwent	18	32	21	0
Gwynedd	5	47	7	0
Mid Glamorgan	2	5	2	0
Powys	19	69	29	0
S. Glamorgan	1	1	1	0
W. Glamorgan	5	13	5	0
Totals	**83**	**275**	**100**	**0**

TABLE 6 - Statutory Varroa Search England & Wales 1993

ENGLAND	No. of Beekeepers	No. of Colonies Examined	No. of Apiaries Sampled	Varroa Positive Apiaries
Avon	12	45	19	8
Bedfordshire	31	260	49	0
Berkshire	34	124	43	1
Buckinghamshire	9	61	12	0
Cambridgeshire	36	223	39	1
Cheshire	28	306	38	0
Cleveland	29	102	31	0
Cornwall	8	12	8	0
Cumbria	52	221	58	0
Derbyshire	25	271	36	0
Devon	12	31	14	12
Dorset	10	40	15	8
Durham	29	118	30	0
E. Sussex	17	87	26	0
Essex	55	428	63	11
Gloucestershire	43	321	50	0
Greater London	16	28	19	7
Greater Manchester	12	63	17	0
Hampshire	14	72	16	3
Hereford & Worcester	89	480	116	1
Hertfordshire	52	349	60	20
Humberside	42	363	54	5
Isle of Wight	0	0	0	0
Kent	6	28	7	6
Lancashire	47	152	53	0
Leicestershire	17	199	30	0
Lincolnshire	11	343	21	4
Merseyside	13	44	13	0
Norfolk	27	415	40	2
N. Yorkshire	89	380	94	0
Northamptonshire	26	320	49	0
Northumberland	50	257	56	0
Nottinghamshire	44	425	59	0
Oxfordshire	23	118	34	0
Shropshire	60	383	81	1
Somerset	30	81	38	22
S. Yorkshire	63	304	78	0
Staffordshire	28	248	53	0
Suffolk	52	444	63	15
Surrey	26	54	30	21
Tyne & Wear	17	58	18	0
Warwickshire	11	171	20	0
West Midlands	15	173	20	0
West Sussex	30	110	34	11
W. Yorkshire	50	250	54	1
Wiltshire	34	178	48	0
Totals	**1,425**	**9,140**	**1,806**	**160**

WALES	No. of Beekeepers	No. of Colonies Examined	No. of Apiaries Sampled	Varroa Positive Apiaries
Clwyd	31	166	31	0
Dyfed	77	491	87	0
Gwent	19	196	26	0
Gwynedd	42	200	47	0
Mid Glamorgan	7	208	14	0
Powys	29	188	39	0
S. Glamorgan	4	21	4	0
W. Glamorgan	8	31	8	0
Totals	**217**	**1,501**	**256**	**0**

TABLE 7

Examination under the Importation of Bees Order 1980 of queen bees imported into England in 1993.

Country of Origin	No. of Queens	No. of Samples Examined	No. of Samples with Nosema	No. of Samples with Acarine	No. of Samples with Amoeba	No. of Samples with no Disease
New Zealand	2,057	44	21	0	0	23
Canary Islands	407	5	3	0	0	2
Hawaii	575	13	6	0	0	7
Eire	26	3	1	0	0	2
Totals	3,065	65	31	0	0	34

USEFUL TABLES

BEEKEEPING METRIC CONVERSION TABLES

MILLIMETRES/INCHES

mm	Inch	mm	Inch	mm	Inch
1	$1/25$	43	$1^{11}/_{16}$	360	$10\ ^1/_4$
2	$1/12$	48	$1\ ^9/_{20}$	286	$11\ ^1/_4$
3	$^1/_8$	51	2	292	$11\ ^1/_2$
5	$^3/_{16}$	76	3	298	$11\ ^3/_4$
6	$^1/_4$	108	$4\ ^1/_4$	305	12
8	$^5/_{16}$	114	$4\ ^1/_2$	356	14
9	$^3/_8$	121	$4\ ^3/_4$	413	$16\ ^1/_4$
12.5	$^1/_2$	140	$5\ ^1/_2$	419	$16\ ^1/_2$
16	$^5/_8$	146	$5\ ^3/_4$	431	17
18	$^3/_4$	152	6	448	$17\ ^5/_8$
22	$^7/_8$	159	$6\ ^1/_4$	460	$18\ ^1/_8$
25	1	216	$8\ ^1/_4$	483	$18\ ^1/_4$
27	$1\ ^1/_{16}$	223	$8\ ^3/_4$	508	20
35	$1\ ^3/_8$	232	$9\ ^1/_8$	546	$21\ ^1/_2$
37	$1\ ^9/_{20}$	239	$9\ ^3/_8$	552	$21\ ^3/_4$
38	$1\ ^1/_2$	243	$9\ ^9/_{16}$	559	22
42	$1\ ^5/_8$	254	10		

CENTIGRADE/FAHRENHEIT

° Cent	° Fahr	° Cent	° Fahr
0	32	49	120
5	40	54	130
7	44	60	140
30	86	62	144
34	92	82	180
38	100	90	194
43	110	100	212

BEEHIVE MEASUREMENTS

BOTTOM BEE-SPACE HIVES

Hive	No. of Frames	Frame Size (mm)	Frame Spacing (mm)	Lug Length (mm)	No. of cells in brood Box
National:					
Brood	11	356 x 216	37	38	58000
Super	10	356 x 140	42	38	36000
Modified Commercial:					
Brood	11	406 x 254	37	16	75000
Super	10	406 x 152	42	16	

TOP BEE-SPACE HIVES

Hive	No. of Frames	Frame Size (mm)	Frame Spacing (mm)	Lug Length (mm)	No. of cells in brood Box
Smith:					
Brood	11	356 x 216	37	18	58000
Super	10	356 x 140	42	18	36000
Langstroth:					
Brood	10	448 x 232	35	16	68000
Super	10	448 x 140	35	16	
Jumbo:					
Brood	10	448 x 286	35	16	85000
Super	10	448 x 140	35	16	
Modified Dadant:					
Brood	11	448 x 286	37	16	93000
Super	10	448 x 159	42	16	

CONVERSION FACTORS

Temperature:	
Fahrenheit to Celcius (Centigrade)	Subtract 32, multiply by .5555 ($^5/_9$)
Celcius to Fahrenheit	Multiply by 1.8 ($^9/_5$) and add 32
Weight:	
Ounces to grams	Multiply by 28.3495
Pounds to grams	Multiply by 453.59237
Hundredweights to kilograms	Multiply by 50.8
Grams to ounces	Divide by 28.3495
Kilograms to pounds	Multiply by 2.2142
Length:	
Inches to centimetres	Multiply by 2.54
Yards to metres	Multiply by 0.9144
Miles to kilometres	Multiply by 1.609
Centimetres to inches	Multiply by 0.3937
Metres to yards	Multiply by 1.0936
Kilometres to miles	Divide by 1.609
Area:	
Acres to hectares	Multiply by 0.404686
Hectares to acres	Multiply by 2.47105
Volume:	
Pints to litres	Multiply by 0.5683
Gallons to litres	Multiply by 4.546
Litres to pints	Multiply by 1.7598
Litres to gallons	Multiply by 0.21997

INTERNATIONAL QUEEN MARKING COLOURS ARE:

YEAR ENDING	1 AND 6	WHITE	WILL
YEAR ENDING	2 AND 7	YELLOW	YOU
YEAR ENDING	3 AND 8	RED	RAISE
YEAR ENDING	4 AND 9	GREEN	GOOD
YEAR ENDING	5 AND 0	BLUE	BEES ?

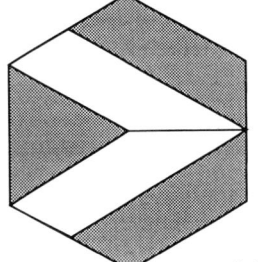

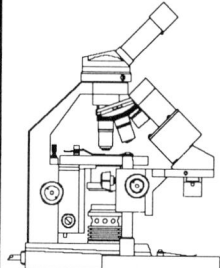

NORTHERN BEE BOOKS

QUEENS' LAND

Norman Rice
132 pages, S/B, £10.95

The most exciting real life story in beekeeping. How Norman Rice grew from having one colony to exporting 30,000 queens per year. Packed with handy hints from a practical beekeeper.

THE BEE BOOK BOOK

Geoffrey Lawes
108 pages, S/B, £9.25.
H/B with slip-case £19.25.

A quite excellent and very readable account of the 'how','why' and 'worth' of collecting bee books. It deals with all aspects of book collection in terms of apicultural literature. **Essential reading for all collectors.**

BEAUTIFUL QUEENS AND HONEY TOO!

Brierley Field's
28 pages, S/B, £1.95

The subtitle - **Sladen's Techniques for Queen Rearing Updated** - explains the purpose of this pamphlet. It describes how the priciples employed by Sladen and used by Brierley Field's with modern techniques and equipment can be used by beekeepers for the production of quality queens. Useful to all those who keep bees as a hobby or on a larger scale.

BEEKEEPING AT BUCKFAST ABBEY

Brother Adam.
122 pages, S/B, £8.45

In this unique record, Brother Adam looks back over a long tradition of beekeeping at Buckfast - a tradition of which he is now an integral part.

SKEPS - Their History, Making and Use.

Frank Alston
105 pages, S/B, £7.75.

The history of skep beekeeping. With clearly explained sections on skep making techniques, selection of materials and the construction of shelters and bee boles. **"The best book on the subject".**

HONEY FARMING

R.O.B.Manley
230 pages, S/B, £8.95

"Honey Farming" was not written for the novice but every beekeeper can learn from it. Too revolutionary to please conservative beekeepers when it was first published 50 years ago "Honey Farming" will delight the modern reader with the author's originality of mind and richness of experience. Any beekeeper who is not familiar with the work of R.O.B. Manley is depriving himself of a genuine and lasting pleasure.

See *The Bee Book Paper* for reviews and a full list of books, slides, videos etc. available

EVOLUTION, RELIGION AND THE UNKNOWN GOD

Georges Van Vrekhem